Limits

A Transition to Calculus

O. LEXTON BUCHANAN, JR.

Albert E. Meder, Jr., EDITORIAL ADVISER

HOUGHTON MIFFLIN COMPANY • BOSTON
Atlanta · Dallas · Geneva, Ill. · Hopewell, N.J. · Palo Alto

ABOUT THE AUTHOR

O. Lexton Buchanan, Jr., Chairman of the Department of Mathematics, Sandy Springs High School, Atlanta, Georgia. Dr. Buchanan has been active in many areas of mathematics education, including writing projects, state mathematics organizations, and teacher training. Having taught mathematics at several high schools and universities, he appreciates the need for a careful development of the precalculus topic of limits.

EDITORIAL ADVISER

Albert E. Meder, Jr., Professor of Mathematics and Dean of the University, Emeritus, Rutgers University. Dr. Meder was Executive Director of the Commission on Mathematics of the College Entrance Examination Board, and he has been an advisory member of the School Mathematics Study Group.

 Printed in the U.S.A.

ISBN: 0-395-17941-6

Contents

1 • *Introduction to Sequences*

1. Introductory Remarks

As you stɪ'dy the ideas set forth in this book, you will come face-to-face with what might well be proclaimed the most dynamic concept in all of mathematics: the concept of a *limit*.

The idea of a limit is a relatively simple idea, but it is not an isolated idea. It cannot be viewed effectively without reference to some of the mathematical notions to which it is firmly attached. Your present study of limits will use sequences of numbers as a major reference point, with inequalities and subsets of real numbers as tools.

Basic to the massive accumulation of practical applications of calculus is the notion of a limit. An understanding of limits is also central to a mature understanding of the real numbers and to the development of formulas for areas and volumes of geometrical figures. Certain applications of limits are mentioned in this section, in hopes of whetting your appetite for some of the exciting encounters to follow throughout your study of limits and of calculus.

Problem A missile is fired at a target 64 miles away. The distance in miles of the missile from its starting point at the end of t minutes is given by the function $f(t) = \frac{1}{4}t^2$. Determine the speed of the missile at the precise instant when it strikes the target.

Solution: Since $f(16) = \frac{1}{4}(16)^2 = 64$, the missile reaches its destination in exactly 16 minutes. The *average speed* during that time interval is given by the formula $R = \frac{D}{T}\left(\text{Rate} = \frac{\text{Distance}}{\text{Time}}\right)$ and is $\frac{64}{16} =$ 4 *miles per minute*. The formula $R = \frac{D}{T}$ yields the average rate of speed in any specified time interval. For example, during the last 8 minutes of the flight the distance traveled is $f(16) - f(8)$ $= 64 - 16 = 48$, so $R = \frac{48}{8} = 6$ *miles per minute*.

Still we are faced with the problem of determining the rate of speed of the missile at the precise instant when $t = 16$. To do so, the simple formula $R = \frac{D}{T}$ seems inadequate. Moreover, we are not even sure yet what is meant by the notion of speed of a moving object at a precise moment — unless the object is moving at a constant rate of speed. But this doesn't appear to be true of the missile, for apparently its speed is constantly increasing.

Instead of waiting for someone to issue us a magic formula suitable for determining the speed of the missile at the exact instant $t = 16$, let us set out to approximate the answer at least. It often happens in mathematics that by making successive approximations to a solution, the true answer becomes readily apparent.

Since the speed of the missile seems to be increasing as t varies from 0 to 16 minutes, we can best proceed by directing attention to time intervals close to the end of the flight. We shall use the formula $R = \frac{D}{T}$ to determine the average rate of speed for smaller and smaller time intervals at the end of the flight. To simplify the writing, we shall use the notation $R_{[a,16]}$ to mean "the average speed of the missile as t varies from a minutes to 16 minutes." For example, $R_{[8,16]}$ is the average speed as the time varies from 8 to 16 minutes and is $\frac{48}{8} = 6$ miles per minute as already shown.

The following values for a will be used: 0, 8, 12, 14, 15, $15\frac{1}{2}$, $15\frac{3}{4}$.

$$R_{[0,16]} = \frac{64}{16} = 4 \text{ miles per minute,}$$

$$R_{[8,16]} = \frac{48}{8} = 6 \text{ miles per minute,}$$

$$R_{[12,16]} = \frac{28}{4} = 7 \text{ miles per minute,}$$

$$R_{[14,16]} = \frac{15}{2} = 7\tfrac{1}{2} \text{ miles per minute,}$$

$$R_{[15,16]} = \frac{7\frac{3}{4}}{1} = 7\tfrac{3}{4} \text{ miles per minute,}$$

$$R_{[15\frac{1}{2},16]} = \frac{3\frac{15}{16}}{\frac{1}{2}} = 7\tfrac{7}{8} \text{ miles per minute,}$$

$$R_{[15\frac{3}{4},16]} = \frac{1\frac{63}{64}}{\frac{1}{4}} = 7\tfrac{15}{16} \text{ miles per minute.}$$

Thus, as we select smaller and smaller time intervals during the latter portion of the flight, we obtain a sequence of corresponding rates of speed. The first seven terms of this sequence are 4, 6, 7, $7\frac{1}{2}$, $7\frac{3}{4}$, $7\frac{7}{8}$ and $7\frac{15}{16}$ miles per minute. It appears that if we were to continue calculating terms of this sequence, using smaller and smaller time intervals, we would come *closer and closer to the number 8* and would therefore have reason to regard 8 miles per minute as the speed of the missile at the precise moment when $t = 16$. It *is* true that we would come closer and closer to 8, as we will show in later portions of this book.

The above problem involves the notion of a sequence and the notion of limit of a sequence. It should suggest to you some of the ideas which we will pursue in this book.

For example, using basic ideas of mensuration from geometry, we can determine areas of triangles, rectangles, and other polygons by simple formulas. However, the problem of determining area is another matter altogether if one of the boundaries of the region is a *curve*. One way to proceed is to approximate such an area by "almost" filling it with rectangles, the sum of whose areas can be found using elementary geometry. As suggested in Figure 1, the area of the region bounded by the x-axis, the y-axis, the vertical line $x = 3$ and the curve $y = (x - 2)^2 + 1 = x^2 - 4x + 5$ can be approximated by inscribing 10 rectangles of equal width and obtaining the sum of their areas. If we were to increase the number of rectangles, say to 20 (again of equal width), a better approximation of the area would be found. By using 30 rectangles, 40 rectangles, 50 rectangles, etc., we would obtain better and better approximations, and the numbers obtained would get closer and closer to some particular number, although they would never actually reach it. We call this number the limit of the approximations, and it represents the actual area of the region.

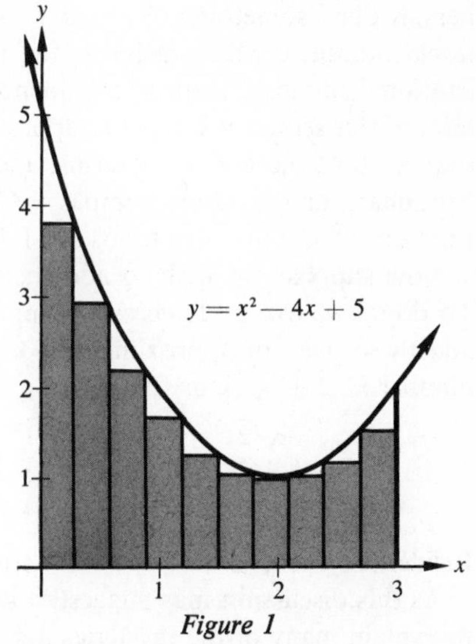

Figure 1

Irrational Numbers

The first problem of this section pointed out that we can regard a new idea, that of speed of a moving object at a precise instant, in terms of an old idea, namely, the formula $R = \frac{D}{T}$, using the concept of limit to bridge the gap between the two ideas. The preceding example pointed out that we can use an old idea, namely the formula for the area of a rectangle, to work out a new idea, a method for finding the area of a region, one of whose boundaries is a curve. Again the accompanying reasoning must involve the notion of limit.

The matter of relating "something old" — i. e., principles and techniques already known — to "something new" is a major characteristic of mathematical reasoning. To understand mathematics, one must realize that more is involved than just using formulas. Some thinking must take place, often involving the adaptation of old ideas to new ones.

To illustrate further, let us assume familiarity with the decimal fractions (numbers which can be placed in finite decimal form) and use this set of numbers as our "something old." Our "something new" will be the set of irrational numbers, which includes, for example, $\sqrt{2}$, $\sqrt{3}$, $\sqrt[3]{7}$, π. For every irrational number, there is a sequence of decimal fractions such that every term of this sequence is a better approximation of the given irrational number than each of the preceding terms, and the limit of the sequence is the given irrational number. For example, $\sqrt{2}$ is the limit of a sequence of decimal fractions whose first five terms are 1.4, 1.41, 1.414, 1.4142, 1.41421.

Now suppose we wish to add the two irrational numbers $\sqrt{2}$ and $\sqrt{5}$. To do so we can find successive approximations to both $\sqrt{2}$ and $\sqrt{5}$ and add these pairs of approximations to find successive approximations to the number $\sqrt{2} + \sqrt{5}$, as follows:

$\sqrt{2}$:	1.4	1.41	1.414	1.4142	1.41421
$\sqrt{5}$:	2.2	2.23	2.236	2.2360	2.23607
$\sqrt{2} + \sqrt{5}$:	3.6	3.64	3.650	3.6502	3.65028

Incidentally, is $\sqrt{2} + \sqrt{5} = \sqrt{7}$ true? Why or why not?

As this discussion may suggest, a knowledge of the limit concept can help to explain many of the mysteries associated with our number systems, particularly the real number system.

Other Questions Relating to Limits

We shall investigate a number of other general questions in terms of limits, as we proceed through the booklet. A few of these questions are:

How can we find the volume of a right circular cone? of a pyramid? of a sphere?

How are tables of logarithmic functions and trigonometric functions derived? What are natural logarithms?

Historical Note. Some of the basic ideas associated with limits were developed by early mathematicians, such as Euclid (around 300 B.C.) and

Archimedes (287–212 B.C.), from a desire to solve practical problems dealing with area, volume, motion and astronomy.

However, it wasn't until the seventeenth century that mathematicians undertook the serious study of limits that was to become the forerunner of our modern theory of limits. They were concerned with problems relating to sound, light, projectiles, the pendulum, the planet, and the like. Indeed, the problem of defining and calculating instantaneous rates (for example, speed or acceleration of a moving object at a precise instant) concerned almost all seventeenth-century mathematicians.

Although our present knowledge of the theory of limits was first envisioned by early mathematicians, it was then neither stated nor understood with present-day clarity. Vagueness prevailed and caused a great deal of suspicion of the "new mathematics" of the day. Numerous debates and conflicts arose regarding this new subject. Not until the early part of the nineteenth century was a clear, concise definition of limit formulated. Since then, mathematicians have organized the ideas of calculus into a cohesive structure, which is closely entwined with the general notion of limit.

The remainder of this chapter will be concerned with describing sequences. In the next chapter, we will delve into the meaning of *limit of a sequence.*

2. What is a Sequence?

The set of *natural numbers* consists of the set of positive integers, which, as you know, is an ordered set, capable of being arranged in order of increasing magnitude (i.e., $\{1, 2, 3, \cdots\}$). This property of the real numbers makes them useful as a reference set in describing sequences. Your present study of limits will center around the idea of a sequence.

If to each natural number n we let correspond one and only one real number, in some specified way, then we say we have devised a *sequence* of numbers. Thus a sequence is really a *set of ordered pairs* having each natural number as a first element once and only once. For example, $\{(1, 1), (2, 3), (3, 5), \cdots, (n, 2n - 1), \cdots\}$ and $\{(1, 1), (2, 4), (3, 9), \cdots, (n, n^2), \cdots\}$ and $\{(1, 2), (2, 2), (3, 0), (4, -4), (5, -10), \cdots, (n, 3n - n^2), \cdots\}$ are sequences. The second element in each ordered pair is called a *term* of the sequence.

To designate a sequence, we select a letter of the alphabet, such as b, and let b_1 (read "b sub 1") designate the first term of the sequence, b_2 the second term, b_3 the third term, etc.: $\{(1, b_1), (2, b_2), (3, b_3), \cdots, (n, b_n), \cdots\}$, or, simply, $\{(n, b_n)\}$. For convenience, we abbreviate the notation $\{(n, b_n)\}$ to $\{b_n\}$ (read "the sequence b sub n"). In other words, when we write $\{b_n\} = \{3n\}$, we are designating the sequence whose nth term is $3n$: $\{(1, 3), (2, 6), (3, 9), \cdots, (n, 3n), \cdots\}$.

From the above discussion we see that a sequence, say $\{b_n\}$, is actually a function whose domain is the set of natural numbers, $\{1, 2, 3, \cdots, n, \cdots\}$, and whose range is the set of terms of the sequence, $\{b_1, b_2, b_3, \cdots, b_n, \cdots\}$. Thus we have the following definition:

DEFINITION

A *sequence* $\{b_n\}$ is a function which associates with every natural number n one and only one real number b_n. The numbers b_n are called the *terms* of the sequence.

Examples **(1)** Suppose to every natural number n we let correspond the number $2n - 5$. Then the first term of the resulting sequence is $2(1) - 5 = -3$, the second term is $2(2) - 5 = -1$, the third term is $2(3) - 5 = 1$, etc. The first few terms of this sequence are displayed as follows:

Natural numbers:	1	2	3	4	5	6	7	$\cdots$
	$\updownarrow$	$\updownarrow$	$\updownarrow$	$\updownarrow$	$\updownarrow$	$\updownarrow$	$\updownarrow$	
Terms of the sequence:	-3	-1	1	3	5	7	9	$\cdots$

(2) Let each natural number n correspond to the number $n^2 - 1$.

Natural numbers:	1	2	3	4	5	6	7	$\cdots$
	$\updownarrow$	$\updownarrow$	$\updownarrow$	$\updownarrow$	$\updownarrow$	$\updownarrow$	$\updownarrow$	
Terms of the sequence:	0	3	8	15	24	35	48	$\cdots$

(3) Let the first five terms be -1, 2, 2, 3, and 4, and the remaining terms be given by the expression $2n$.

Natural numbers:	1	2	3	4	5	6	7	$\cdots$
	$\updownarrow$	$\updownarrow$	$\updownarrow$	$\updownarrow$	$\updownarrow$	$\updownarrow$	$\updownarrow$	
Terms of the sequence:	-1	2	2	3	4	12	14	$\cdots$

(4) Let each natural number n correspond to the number $\sin \frac{n}{2}\pi$.

Natural numbers:	1	2	3	4	5	6	7	8	$\cdots$
	$\updownarrow$	$\updownarrow$	$\updownarrow$	$\updownarrow$	$\updownarrow$	$\updownarrow$	$\updownarrow$	$\updownarrow$	
Terms of the sequence:	1	0	-1	0	1	0	-1	0	$\cdots$

This is an example of a sequence in which more than one term is the same number. The first, fifth, ninth, etc., terms are the number 1, the second, fourth, sixth, eighth, etc., terms are the number 0, and the third, seventh, eleventh, etc., terms are the number -1. Thus we see that repetition of terms is allowable.

(5) If $\{a_n\} = \left\{7 - \frac{1}{n^2}\right\}$, then $a_1 = 6$, $a_2 = 6\frac{3}{4}$, $a_3 = 6\frac{8}{9}$.

In devising and describing sequences, we must make certain that every single term of the sequence can be found from our description. One way to describe a sequence is by a general term involving "n," as in (1), (2), (4), and (5) above; this way will be used most frequently in this booklet. Another way to describe a sequence is by a verbal description, such as: Let the nth term of the sequence be the largest integer whose square is less than or equal to n. The first eleven terms of this sequence are 1, 1, 1, 2, 2, 2, 2, 2, 3, 3, 3.

EXERCISES

For each sequence in the left-hand column, there is a list of numbers in the right-hand column consisting of the first five terms of that sequence. Decide which list of numbers corresponds to which sequence. There are more lists of numbers in the right-hand column than are needed.

Sequences

1. $\{a_n\} = \{2^n - n^2\}$

2. $\{b_n\} = \left\{\sin \frac{n}{2}\pi + \cos \frac{n}{2}\pi\right\}$

3. $\{c_n\} = \{100(.1)^n\}$

4. $\{d_n\} = \{1 + (-1)^n\}$

5. $\{f_n\} = \left\{\frac{n}{2n - 1}\right\}$

6. $\{g_n\} = \left\{\frac{n + 2}{2n + 1}\right\}$

7. $\{h_n\} = \left\{\frac{n + 1}{2n + 1}\right\}$

First Five Terms

(a) 10, 1, .1, .01, .001

(b) 1, 0, -1, 1, 1

(c) 1, 0, -1, 0, 7

(d) 1, -1, -1, 1, 1

(e) 0, 2, 0, 2, 0

(f) 1, $\frac{2}{3}$, $\frac{3}{5}$, $\frac{4}{7}$, $\frac{5}{9}$

(g) $\frac{2}{3}$, $\frac{4}{5}$, $\frac{6}{7}$, $\frac{8}{9}$, $\frac{10}{11}$

(h) 1, $\frac{4}{5}$, $\frac{5}{7}$, $\frac{2}{3}$, $\frac{7}{11}$

(i) $\frac{2}{3}$, $\frac{3}{5}$, $\frac{4}{7}$, $\frac{5}{9}$, $\frac{6}{11}$

3. Describing and Graphing Sequences

Graphing Sequences

The terms of a sequence can be represented by points on the real number line. In Figure 2, the first seven terms of $\{b_n\} = \left\{\frac{1}{n}\right\}$ have been located, and

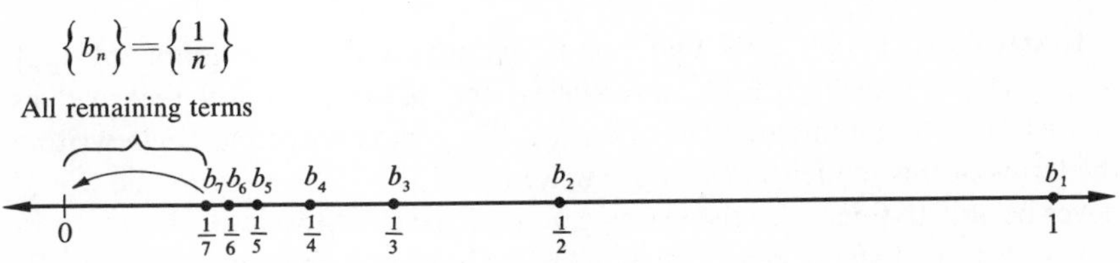

Figure 2

the curved arrow pointing from $\frac{1}{7}$ toward 0 indicates that all remaining terms are between 0 and $\frac{1}{7}$. The fact that the first seven terms were located was purely an arbitrary choice. We could have located as few as three or four terms without sacrificing the clarity of our graph.

The important feature of a graph is that it should enable the reader to visualize where all the points would appear if we were to represent every single term of the sequence by a point on the real number line. Hence the number of points to locate will vary from sequence to sequence and from person to person.

Additional examples of graphs of sequences are shown in Figure 3.

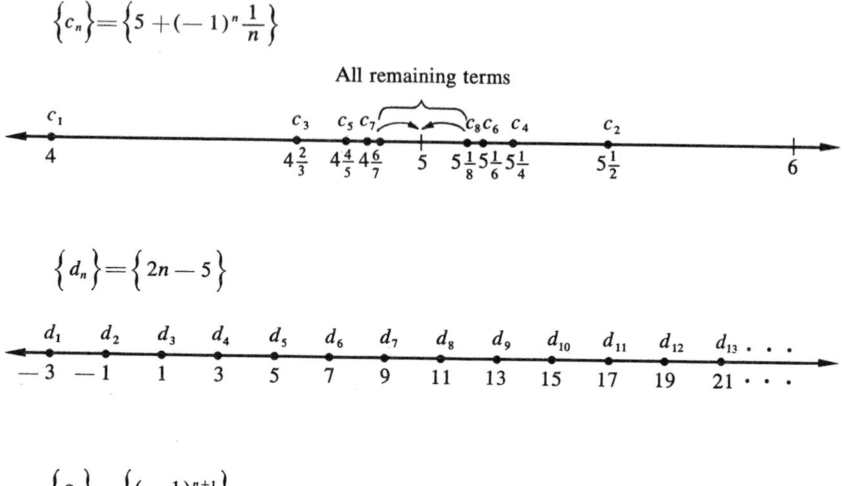

Figure 3

REMARK: Although only two points are involved in the graph of $\{g_n\} = \{(-1)^{n+1}\}$ (see Figure 3), nevertheless the sequence is still regarded as consisting of an infinite number of terms. If a person wished to begin writing the terms of this sequence, he would write $1, -1, 1, -1, 1, -1$, and he would never be able to finish, for the listing goes on forever. Moreover, it is interesting to realize that the sequence $\{(-1)^n\}$, whose first few terms are $-1, 1, -1, 1$, is a different sequence from $\{(-1)^{n+1}\}$.

Additional Ways of Describing Sequences

A sequence can have one expression for certain values of n and another expression for other values of n, that is, a *multiple description.* For example,

$$g_n = \begin{cases} n \text{ for } n \text{ even} \\ \dfrac{1}{n} \text{ for } n \text{ odd}, \end{cases} \quad p_n = \begin{cases} 2^n \text{ for } n = 1, 2, 3, 4 \\ 3^n \text{ for } n \geq 5, \end{cases} \quad r_n = \begin{cases} \dfrac{n}{3} \text{ for } n \text{ a multiple of } 3 \\ -n \text{ otherwise.} \end{cases}$$

A *constant sequence* is a sequence having the same number for every term. To describe a constant sequence $\{a_n\} = c, c, c, \cdots$, where c is any real number, we write $\{a_n\} = \{c\}$. For example, $\{a_n\} = \{2\}$ denotes the constant sequence whose terms are $2, 2, 2, 2, 2, \cdots$.

EXERCISES

1. Graph each of the five sequences whose general description is given below.

$$a_n = \frac{2n}{n+1} \qquad d_n = \begin{cases} \dfrac{1-5n}{n} \text{ for } n \text{ odd} \\ \dfrac{1+5n}{n} \text{ for } n \text{ even} \end{cases}$$

$$b_n = \frac{2^{n+1}+1}{3^n}$$

$$c_n = \frac{n+(-1)^n}{n+1} \qquad f_n = \frac{(-1)^{n+1}}{n+1}$$

2. Find the first five terms of each sequence whose general description is given here.

$$a_n = (-1)^{n+1} \qquad k_n = \begin{cases} 2^n \text{ for } n \text{ odd} \\ \dfrac{1}{2^{n-1}} \text{ for } n \text{ even} \end{cases}$$

$$b_n = (-1)^n$$

$$c_n = \frac{1}{n+2} \qquad p_n = (-2)^n$$

$$d_n = \frac{4}{n+2} \qquad q_n = (-1)^n + 1$$

$$e_n = \frac{1}{n^2+2} \qquad r_n = 1 + \frac{1}{n+1}$$

$$f_n = n(n-1) \qquad s_n = \frac{1}{n^3 - 6(n-1)^2}$$

$$h_n = \frac{1}{n}(-1)^n$$

$$j_n = \begin{cases} n+1 \text{ for } n \text{ odd} \\ n-1 \text{ for } n \text{ even} \end{cases} \qquad t_n = \begin{cases} n \text{ for } 1 \leq n \leq 7 \\ 8 \text{ for } n \geq 8 \end{cases}$$

3. Sometimes a sequence can be described in more than one way. For instance, $\{d_n\} = \{(-1)^n\}$ can also be described by the multiple description

$$d_n = \begin{cases} -1 \text{ for } n \text{ odd} \\ 1 \text{ for } n \text{ even.} \end{cases}$$

(Cont.)

Each of the sequences whose description is given below is the same as some sequence in Exercise 2. Match each of the descriptions below with its corresponding sequence in Exercise 2.

(a) $n + (-1)^{n+1}$

(b) $\begin{cases} 1 \text{ for } n \text{ odd} \\ -1 \text{ for } n \text{ even} \end{cases}$

(c) $\begin{cases} 0 \text{ for } n \text{ odd} \\ 2 \text{ for } n \text{ even} \end{cases}$

(d) $\dfrac{1}{n + (n-1)(n-2)(n-3)}$

(e) $\dfrac{n+2}{n+1}$

(f) $\begin{cases} -\dfrac{1}{n} \text{ for } n \text{ odd} \\ \dfrac{1}{n} \text{ for } n \text{ even} \end{cases}$

(g) $(-1)^n 2^n$

4. Similarities Among Sequences

A question such as the following one is sometimes encountered: "Find the sixth term in a sequence whose first five terms are $\frac{1}{1}, \frac{4}{3}, \frac{9}{5}, \frac{16}{7}$, and $\frac{25}{9}$."

There would be little disagreement that the *most likely* sixth term is $\frac{36}{11}$, with the general term being $\frac{n^2}{2n-1}$. It is interesting to note that there is no absolute requirement for the general term to be $\frac{n^2}{2n-1}$ for all remaining terms of the sequence. We could have this general term for the first five terms and some other description for the remaining terms, or for some of the remaining terms. Moreover, there is a general term which yields the five given terms and a sixth term which is different from $\frac{36}{11}$:

$$\{a_n\} = \left\{\frac{n^2 + (n-1)(n-2)(n-3)(n-4)(n-5)}{2n-1}\right\}.$$

Using this general term, $a_6 = \frac{156}{11}$.

Thus the question presented at the beginning of this section does not have a completely unique answer. However, we should realize that the question is a reasonable one in the sense that any general term other than $\frac{n^2}{2n-1}$ would be much less likely than this implied general term. It is common practice to imply the general pattern of a sequence by listing the first few terms, and for many purposes this practice is desirable.

However, since our work with sequences in this book will be extensive — often involving rather unusual and somewhat surprising sequences — we will give a very careful and complete description of each sequence, in order to avoid possible ambiguity.

In this section we shall give attention to other examples of sequences which are alike for the first few terms (or for certain other terms) but which are not alike for all terms.

For example, suppose the first three terms of a sequence are $\frac{3}{2}$, $\frac{5}{3}$, and $\frac{7}{4}$. What would be a likely fourth term for such a sequence? Should the numerator of the fourth term be 9, since the numerators of the first three terms are increasing by 2 each time? Not necessarily. The numerators 3, 5, and 7 are the first three odd prime natural numbers, so that the numerator of the fourth term could be 11, the fourth odd prime natural number. Should the denominator of the fourth term be 5? Even though the denominators of the second and third terms are each one more than the denominator of the preceding term, there is no compelling reason for the denominator of the fourth term to follow the same pattern. The fourth, fifth, and sixth denominators could each be *2* more than the preceding one; the seventh, eighth, and ninth denominators could each be *3* more than the preceding one, and so forth for succeeding triples of denominators. Indeed, a clever student could devise all sorts of sequences having the first three terms $\frac{3}{2}$, $\frac{5}{3}$, and $\frac{7}{4}$.

Consider another example. If someone asks, "What term is next in a certain sequence whose first three terms are 1, $\frac{1}{2}$, and $\frac{1}{3}$?" a unique answer cannot be given. It is true that if the sequence is $\left\{\frac{1}{n}\right\}$, the next term is $\frac{1}{4}$, but there are infinitely many sequences with these same first three terms and different terms thereafter. Two such sequences are described by

$$a_n = \frac{1}{n + (n-1)(n-2)(n-3)},$$

whose first five terms are 1, $\frac{1}{2}$, $\frac{1}{3}$, $\frac{1}{10}$, $\frac{1}{29}$, and by

$$b_n = \begin{cases} -\frac{1}{n} \text{ for } n \text{ a multiple of } 4 \\ \frac{1}{n} \text{ otherwise,} \end{cases}$$

whose first eight terms are 1, $\frac{1}{2}$, $\frac{1}{3}$, $-\frac{1}{4}$, $\frac{1}{5}$, $\frac{1}{6}$, $\frac{1}{7}$, $-\frac{1}{8}$.

The expression $(n-1)(n-2)(n-3)$ gives 0 for $n = 1$, 2, and 3 and therefore causes $\frac{1}{n + (n-1)(n-2)(n-3)}$ to equal $\frac{1}{n}$ for $n = 1$, 2, and 3. This expression and similar expressions — for example, $(n-1)(n-2)$ and $(n-1)(n-2)(n-3)(n-4)$ — can be used in devising sequences having certain specified terms. The first four terms of

$$\{a_n\} = \left\{\frac{n+1}{2n-(n-1)(n-2)(n-3)(n-4)}\right\}$$

are the same numbers as the first four terms of $\{b_n\} = \left\{\frac{n+1}{2n}\right\}$, but $a_5 = -\frac{3}{7}$, whereas $b_5 = \frac{3}{5}$. If we let $\{a_n\}$ equal either

$$\left\{\frac{n+1}{2n+(n-1)(n-2)(n-3)(n-4)}\right\}$$

or

$$\left\{\frac{n+1+(n-1)(n-2)(n-3)(n-4)}{2n}\right\},$$

we would get similar results.

A *multiple description* is often useful for devising a sequence having some, but not all, terms equal to corresponding terms of another sequence. For example,

$$a_n = \begin{cases} \frac{3}{n} \text{ for } 1 \leq n \leq 10 \\ 3n \text{ otherwise,} \end{cases} \qquad b_n = \begin{cases} \frac{2}{n} \text{ for } 1 \leq n \leq 10 \\ 3n \text{ otherwise.} \end{cases}$$

EXERCISES

In Exercises 1–6, which of the sequences listed in (a) through (m) meet the condition described.

1. Three sequences whose first three terms are 5, 10, 17.
2. Two sequences whose first three terms are $0, \frac{1}{5}, \frac{4}{5}$.
3. One sequence whose first four terms are 0, 1, 8, 27.
4. Two sequences whose first three terms are 8, 27, 64.
5. Two sequences whose first two terms are $\frac{1}{2}, \frac{1}{6}$.
6. Three sequences whose first three terms are 1, 3, 7.

(a) $\{a_n\} = \{(n+1)^3\}$

(b) $\{b_n\} = \{(n+1)^2+1\}$

(c) $\{c_n\} = \left\{\frac{(n-1)^2}{5+(n-1)(n-2)(n-3)}\right\}$

(d) $\{d_n\} = \left\{\frac{1}{n^2+n}\right\}$

(e) $\{e_n\} = \left\{\frac{(n+1)^3}{n^3-6n^2+11n-5}\right\}$

(f) $\{f_n\} = \{n^2 + 2n + 2 + (n - 1)(n - 2)(n - 3)\}$

(g) $\{g_n\} = \{[2^{n-1} - \frac{1}{2}][2 + (n - 1)(n - 2)(n - 3)]\}$

(h) $\{h_n\} = \left\{\frac{(n - 1)^2}{5}\right\}$

(i) $\{p_n\} = \{2^n - 1\}$

(j) $\{r_n\}$, where $r_n = \begin{cases} 2^n - 1 & \text{for } n = 1, 2, 3 \\ n^2 & \text{for } n > 3 \end{cases}$

(k) $\{s_n\} = \left\{\frac{2^{(n-1)(n-2)}}{n(n + 1)}\right\}$

(l) $\{t_n\} = \{(n - 1)^3\}$

(m) $\{v_n\} = \{n^2 + 2n + 2 - (n - 1)(n - 2)(n - 3)\}$

Devise a sequence which meets the conditions given.

7. A sequence $\{f_n\}$ whose first four terms are 2, 4, 8, and 16 and whose fifth term is *not* 32.

8. A sequence $\{g_n\}$ such that $g_n = (n - 1)^3$ for $n = 1, 2$, and 3, but $g_4 \neq (4 - 1)^3$.

9. A sequence $\{s_n\}$ whose first four terms are 1, 3, 7, and 15.

For each exercise below define two sequences $\{a_n\}$ and $\{b_n\}$ in which

10. The first four terms are 4, 9, 16, and 25, and the fifth terms differ.

11. The first four terms are 5, 10, 17, and 26, and the fifth terms differ.

5. Recursive Descriptions

The remarkable Italian mathematician, Leonardo of Pisa, who was better known by his nickname Fibonacci (an abbreviation of "filius Bonacci"), wrote a book called *Liber Abacci*, which means "a book about the abacus." The second version of this book appeared in 1228 and contained nearly all of the arithmetical and algebraic knowledge of those times. One of the examples from this book follows.

Someone placed a pair of rabbits in a pen with the objective of finding out how many pairs of rabbits would be born there in the course of a year. The assumptions were (1) that every month a pair of rabbits produces another pair, and (2) that rabbits begin to bear young two months after their own birth.

At the beginning of the first month, there was *1* pair, and this pair was newly-born, so that at the end of the first month there was still *1* pair. Since the first pair produced an offspring in the second month, there were *2* pairs at the end of the second month. Of these, one pair, namely the first pair, gave birth in the following month, so that at the end of the third month there were

3 pairs. Of these, two pairs produced offspring in the following month, so that at the end of the fourth month there were $3 + 2 = 5$ pairs of rabbits. It should be noted that the 3 in $3 + 2 = 5$ represents the number of pairs at the end of the third month and that the 2 in the $3 + 2 = 5$ represents the number of pairs at the end of the second month. At the end of the fifth month, there were $5 + 3 = 8$ pairs, the 5 representing the number of pairs at the end of the fourth month and the 3 representing the number of pairs at the end of the third month.

Continuing this process, we see that

At the end of the	there were
sixth month	$8 + 5 = 13$ pairs
seventh month	$13 + 8 = 21$ pairs
eighth month	$21 + 13 = 34$ pairs
ninth month	$34 + 21 = 55$ pairs
tenth month	$55 + 34 = 89$ pairs
eleventh month	$89 + 55 = 144$ pairs
twelfth month	$144 + 89 = 233$ pairs

Thus the first thirteen terms of the sequence of number of pairs of rabbits are 1, 1, 2, 3, 5, 8, 13, 21, 34, 55, 89, 144, 233. These are the first thirteen terms of the *Fibonacci sequence.*

The Fibonacci sequence is described (without using rabbits) as the sequence $\{a_n\}$ such that $a_1 = 1$, $a_2 = 1$, and for $n \geq 3$, $a_n = a_{n-2} + a_{n-1}$ (i.e., any term after the second term is the sum of the two preceding terms). The Fibonacci sequence is one example of a sequence described recursively.

DEFINITION

A sequence is said to be *described recursively* if the first *R* terms are stated, for some natural number *R*, and each succeeding term is defined as a function of one or more of the preceding terms.

Examples *Sequences Described Recursively*

(1) Let $b_1 = 3$, $b_2 = 5$, and for $n \geq 3$, $b_n = b_{n-2} + b_{n-1} + 1$. The first five terms are 3, 5, 9, 15, 25.

(2) Let $c_1 = 2$, $c_2 = 3$, and for $n \geq 3$, $c_n = (c_{n-1})^2 - c_{n-2}$. Then $c_3 = 3^2 - 2 = 7$, $c_4 = 7^2 - 3 = 46$, $c_5 = 46^2 - 7 = 2109$.

(3) Let $d_1 = 2$, $d_2 = 1$, $d_3 = \frac{2}{3}$, and for $n \geq 4$, $d_n = \left(\frac{1}{d_{n-1}}\right)^2$.

Then $d_4 = \left(\frac{1}{\frac{2}{3}}\right)^2 = \frac{9}{4}$, $d_5 = \left(\frac{1}{\frac{9}{4}}\right)^2 = \frac{16}{81}$.

Many sequences which have been described recursively can also be described by a general term, and vice versa. Some such sequences are discussed in the following set of examples.

Examples **(1)** Let $a_1 = 3$, and for $n \geq 2$, $a_n = (a_{n-1})^2$. Then the first five terms are 3, 3^2, 3^4, 3^8 and 3^{16}. This sequence can also be described by the general term $3^{2^{n-1}}$.

(2) Let $b_1 = 2$, and for $n \geq 2$, $b_n = 3b_{n-1}$. That is, every term after the first term is 3 times the preceding term. The first five terms are 2, 6, 18, 54 and 162. This sequence can also be described by the general term $2(3^{n-1})$.

The sequence in Example (2) is an example of a *geometric sequence*. A geometric sequence is a sequence in which the ratio of any term (after the first term) to its predecessor is always the same. A geometric sequence can be described by the general term a_1r^{n-1}, where a_1 is the first term and r is the common ratio. The numbers a_1 and r may be positive or negative. The definition tells us that $\frac{a_2}{a_1} = r$, $\frac{a_3}{a_2} = r$, and in general, $\frac{a_{n+1}}{a_n} = r$, for all natural numbers n. Notice that any geometric sequence $\{a_n\} = \{a_1r^{n-1}\}$ can be described recursively as follows: Let a_1 be specified, and for $n \geq 2$ let $a_n = r(a_{n-1})$. Using this description, $\frac{a_2}{a_1} = \frac{a_1r}{a_1} = r$, $\frac{a_3}{a_2} = \frac{a_1r^2}{a_1r} = r$, and so on.

For example, $\{a_n\} = \{4(\frac{1}{2})^{n-1}\}$ is a geometric sequence; the first five terms are 4, 2, 1, $\frac{1}{2}$, and $\frac{1}{4}$. This sequence can be described recursively by stating that $a_1 = 4$, and for $n \geq 2$, $a_n = \frac{1}{2} \cdot a_{n-1}$.

(3) Let $a_1 = 2$, and for $n \geq 2$, $a_n = a_{n-1} + 3$. That is, every term after the first is 3 greater than the preceding term. The first five terms are 2, 5, 8, 11, and 14. This sequence can also be described by the general term $3n - 1$.

The sequence in Example (3) is an example of an *arithmetic sequence*. An arithmetic sequence is a sequence in which the difference between any term (after the first term) and its predecessor is always the same. An arithmetic sequence can be described by the general term $a_1 + d(n - 1)$, where a_1 is the first term and d is the common difference. The numbers a_1 and d may be positive or negative. An arithmetic sequence can also be described recursively by stating the value for a_1 and by letting $a_n = a_{n-1} + d$ for $n \geq 2$.

For example, the sequence $\{a_n\} = \{5n - 7\} = \{-2 + 5(n - 1)\}$ is an arithmetic sequence with a first term of -2 and common difference of 5. Thus its first five terms are -2, 3, 8, 13, and 18. The sequence can be described recursively by stating that $a_1 = -2$, and for $n \geq 2$, $a_n = a_{n-1} + 5$.

EXERCISES

Find the first five terms of each sequence described below.

1. $a_1 = 1, a_2 = 2$; for $n \geq 3$, $a_n =$ the sum of all preceding terms.
2. $b_1 = 1, b_2 = 2$; for $n \geq 3$, $b_n = (b_{n-2})^2 + b_{n-1}$.
3. $c_1 = 1, c_2 = 2, c_3 = 3$; for $n \geq 4$, $c_n = 2c_{n-1}$.
4. $d_1 = 1, d_2 = 2$; for $n \geq 3$, $d_n = (d_{n-1})^2 - 1$.
5. $f_1 = 1, f_2 = 2$; for $n \geq 3$, $f_n = (f_{n-1})^2 - f_{n-2}$.
6. Which sequences in Exercises 1–5 are the same?
7. Find the first four terms of $c_n = \dfrac{(1 + \sqrt{5})^n - (1 - \sqrt{5})^n}{2^n \sqrt{5}}$. (The answer may be very surprising!)
8. Let $\{b_n\} = \{n!\}$, where $n! = (1)(2)(3)\cdots(n)$. Then $b_1 = 1$, $b_2 = (1)(2) = 2$, $b_3 = (1)(2)(3) = 6$, etc. Describe $\{b_n\}$ recursively by specifying that $b_1 = 1$ and stating a general expression for b_n in terms of b_{n-1} for $n \geq 2$.

For each of the following sequences, find a general term which describes the sequence:

9. $\{a_n\}$: $a_1 = 3$; for $n \geq 2$, $a_n = a_{n-1} + 12$.
10. $\{b_n\}$: $b_1 = 1$; for $n \geq 2$, $b_n = -b_{n-1}$.
11. $\{c_n\}$: $c_1 = 5$; for $n \geq 2$, $c_n = -c_{n-1}$.
12. $\{d_n\}$: $d_1 = 2$; for $n \geq 2$, $d_n = \frac{1}{3} d_{n-1}$.
13. $\{f_n\}$: $f_1 = -5$; for $n \geq 2$, $f_n = f_{n-1} - 2$.

Describe each of the following sequences recursively.

14. $\{a_n\} = \{3n + 5\}$. **15.** $\{b_n\} = \{3^n\}$. **16.** $\{c_n\} = \{\log_{10} 2^n\}$.
17. Given that the first two terms of an arithmetic sequence $\{a_n\}$ are p and q, what is the third term, in terms of p and q? the general term?
18. Which terms of $\{b_n\} = \{1 + 4(n - 1)\}$ are also terms of $\{c_n\} = \{-1 + 6(n - 1)\}$? Devise a sequence whose terms are these common terms. Is this sequence an arithmetic sequence?
19. Let $\{a_n\}$ be the Fibonacci sequence. The first seven terms are 1, 1, 2, 3, 5, 8, 13. Notice that:

a_1 is 1 less than a_3,
$a_1 + a_2 = 2$ is 1 less than a_4,
$a_1 + a_2 + a_3 = 4$ is 1 less than a_5,
$a_1 + a_2 + a_3 + a_4 = 7$ is 1 less than a_6.

In general, $a_1 + a_2 + a_3 + \cdots + a_n = a_{n+2} - 1$.
Show that this generalization is true. [Hint: In the general expression, replace a_1 with $a_3 - a_2$, replace a_2 with $a_4 - a_3$, and so forth (since $a_1 + a_2 = a_3$, $a_2 + a_3 = a_4$, etc.). Or you may wish to show this by mathematical induction.]
20. As in Problem 19, let $\{a_n\}$ be the Fibonacci sequence. Then

$a_1 + a_3$ equals which term?
$a_1 + a_3 + a_5$ equals which term?
$a_1 + a_3 + a_5 + a_7$ equals which term?

In general, $a_1 + a_3 + a_5 + \cdots + a_{2n-1}$ equals which term? Show why.

2 • *Limit of a Sequence*

1. Intervals on the Real Number Line; Neighborhoods

The graphing of sequences utilizes the concept of a one-to-one correspondence between the real numbers and the points on the real number line. This means that every point on the real number line corresponds to exactly one real number and, vice versa, every real number corresponds to exactly one point on the real number line.

Thus, when discussing sequences, we have a choice of considering the terms *numbers* or *points*. This means that we can work within either an *analytical* setting or a *geometrical setting*. In the analytical setting, the terms of a sequence are referred to as numbers. In the geometrical setting, the terms are referred to as points on the real number line. Sometimes one of the settings is more convenient than the other, so it is very fortunate that we have the two from which to choose.

In mathematical language, certain expressions are analytical and certain expressions are geometrical. The analytical expression "a is less than b" corresponds to the geometrical expression "a is to the left of b." The analytical expression "a is greater than b" corresponds to the geometrical expression "a is to the right of b."

The purpose of this section is to introduce several very important definitions regarding the real numbers and the real number line. *These definitions apply to the real numbers in general and not merely to sequences.*

DEFINITIONS

The notation $[a, b]$, with $a < b$, designates the set of all real numbers x such that $a \leq x \leq b$ (analytical definition) or the set of all points in the line segment whose endpoints are a and b, with a and b included (geometrical definition). Geometrically, $[a, b]$ is called a *closed interval.*

The notation $\langle a, b\rangle$, with $a < b$, designates the set of all real numbers x such that $a < x < b$ (analytical definition) or the set of all points in the line segment whose endpoints are a and b, with a and b excluded (geometrical definition). Geometrically, $\langle a, b\rangle$ is called an *open interval.*

Sometimes it is useful to describe the set of points constituting a line segment which includes one endpoint but not the other endpoint. Such an interval is neither open nor closed.

DEFINITIONS

The notation $\langle a, b]$, with $a < b$, designates the set of all real numbers x such that $a < x \leq b$ (analytical definition) or the set of points $[a, b]$ with a excluded (geometrical definition).

The notation $[a, b\rangle$, with $a < b$, designates the set of all real numbers x such that $a \leq x < b$ (analytical definition) or the set of points $[a, b]$ with b excluded (geometrical definition).

Neighborhoods

Suppose we have a given point, say 5, on the real number line, and we wish to describe those points on the number line whose distance from 5 is less than 1. This would be all numbers x such that $4 < x < 6$. Using interval notation, this would be the open interval $\langle 4, 6\rangle$. This interval can also be referred to as the *neighborhood of 5 of radius 1.* A neighborhood is a cluster of points on the real number line surrounding a particular point on the line. The next definition states this idea.

DEFINITION

For any point r on the real number line, an open interval with r as midpoint is called a *neighborhood of r.* The *radius* of a neighborhood whose endpoints are a and b, with $a < b$, is the number $\frac{b-a}{2}$.

Furthermore, any neighborhood $\langle a, b\rangle$ can be regarded as the intersection of two sets:

(1) the set of all real numbers $> a$, and
(2) the set of all real numbers $< b$.

The compound inequality $a < x < b$ describes this situation; this inequality designates the fact that two conditions must be met: a number x must be $>a$, and it must be $<b$.

In Figure 4, examples are given to illustrate the graphing techniques which we will use to indicate intervals on the real number line. Notice that the graph of a neighborhood is the same as the graph of an open interval, except that the midpoint of the neighborhood must be marked. Any open interval can be considered a neighborhood of its midpoint.

(1) The *open interval* $\langle 2, 3\rangle$.

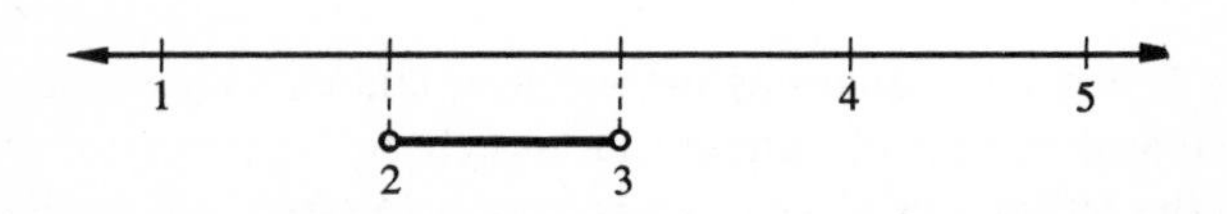

(2) The *neighborhood* $\langle 2, 3\rangle$. (This graph is the same as that in Example 1, except that the midpoint, $2\frac{1}{2}$, is shown.)

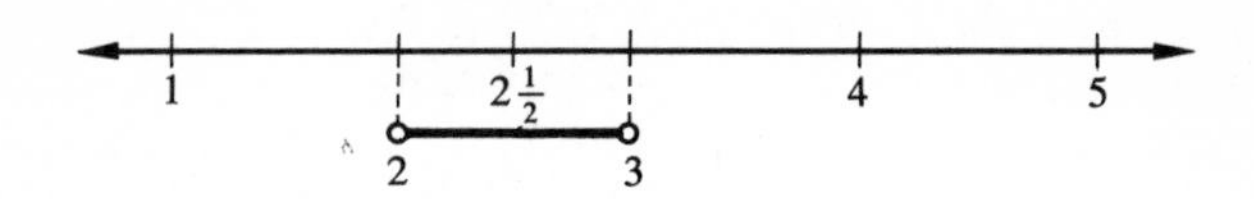

(3) The *closed interval* $[1.5, 3]$.

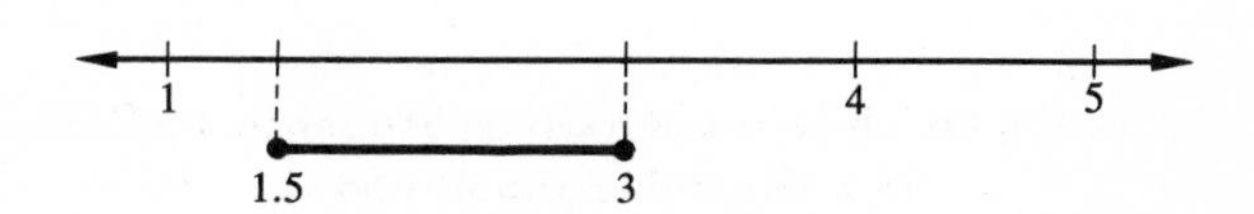

(4) The intervals $[2, 2\frac{3}{4}\rangle$ and $\langle 3\frac{1}{2}, 4\frac{3}{4}]$.

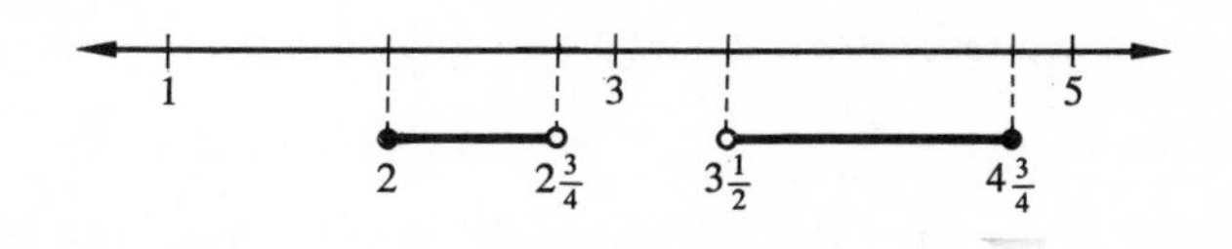

Figure 4

The symbols ∞ ("infinity") and $-\infty$ ("negative infinity") are often employed in the following ways:

DEFINITIONS

The notation $[b, \infty\rangle$ designates the set of all real numbers greater than or equal to b (analytical definition) or the set of all points on a number line to the right of and including b (geometrical definition).

The notation $\langle b, \infty\rangle$ is the set $[b, \infty\rangle$ with b excluded.

The notation $\langle -\infty, b]$ designates the set of all real numbers less than or equal to b (analytical definition) or the set of all points on a number line to the left of and including b (geometrical definition).

The notation $\langle -\infty, b\rangle$ is the set $\langle -\infty, b]$ with b excluded.

REMARK: When two equations or two inequalities have exactly the same solution — that is, the set of numbers for which one of them is true is exactly the same as the set for which the other is true — they are said to be *equivalent.* The symbol $\leftrightarrow$, when placed between two equations or inequalities, is read "is equivalent to." For example, keeping in mind that we are working within the set of natural numbers for n, we can say that

$$\begin{aligned} 2n + 4 \leq 5n + 1 &\leftrightarrow 4 \leq 3n + 1 \\ &\leftrightarrow 3 \leq 3n \\ &\leftrightarrow 1 \leq n. \end{aligned}$$

Also,

$$\begin{aligned} \frac{1}{n} < 5 &\leftrightarrow \left(\frac{n}{5}\right)\left(\frac{1}{n}\right) < \left(\frac{n}{5}\right)(5) \\ &\leftrightarrow \frac{1}{5} < n. \end{aligned}$$

2. Determining Which Terms of a Given Sequence are Contained in a Given Neighborhood

Very shortly we shall explore the meaning of *limit of a sequence.* Before this can be done effectively, however, it is important that certain ideas concerning neighborhoods be discussed. These ideas involve determining which terms of a given sequence are contained in a given neighborhood.

Which terms of $\{a_n\} = \left\{\frac{n+3}{n+10}\right\}$, if any, are contained in the neighborhood $\langle .9, 1.1\rangle$? To answer this question, we need to solve the compound inequality $.9 < \frac{n+3}{n+10} < 1.1$. More specifically, we need to solve the inequality $.9 < \frac{n+3}{n+10}$ (and thereby determine which terms are $> .9$) and the inequality $\frac{n+3}{n+10} < 1.1$ (and thereby determine which terms are < 1.1). The terms which meet *both* of these two conditions will satisfy the compound inequality.

Solving $.9 < \frac{n+3}{n+10}$ gives

$$9 < \frac{10n+30}{n+10} \leftrightarrow 9(n+10) < 10n+30$$
$$\leftrightarrow 9n+90 < 10n+30$$
$$\leftrightarrow 60 < n.$$

Thus all terms a_n with $n > 60$ are $>.9$. Solving $\frac{n+3}{n+10} < 1.1$ gives

$$\frac{10n+30}{n+10} < 11 \leftrightarrow 10n+30 < 11n+110$$
$$\leftrightarrow 0 < n+80,$$

which is true for all natural numbers n. Thus all terms are <1.1. Therefore the neighborhood $\langle .9, 1.1\rangle$ contains all terms a_n with $n > 60$ — that is, a_{61} and all succeeding terms. (See Figure 5.)

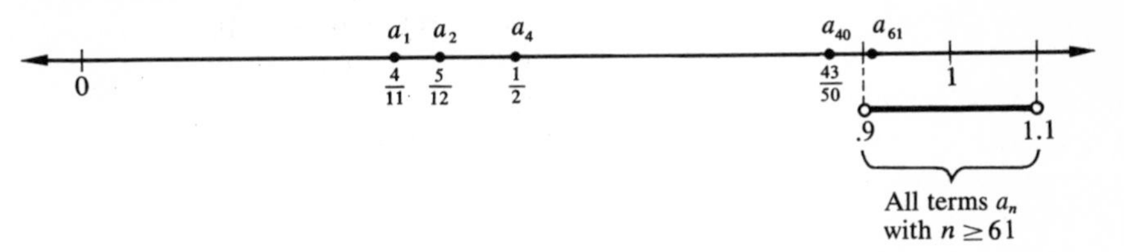

Figure 5. Specifying Which Terms of $\{a_n\} = \left\{\frac{n+3}{n+10}\right\}$ *Are in the Neighborhood* $\langle .9, 1.1\rangle$

As another example, let us determine which terms of $\{b_n\} = \left\{\frac{2n}{n+5}\right\}$ are in the neighborhood $\langle 1, 1.5\rangle$, that is, the neighborhood of 1.25 with radius .25.

The inequality $1 < \frac{2n}{n+5} < 1.5$ must be solved:

$$1 < \frac{2n}{n+5} \leftrightarrow n+5 < 2n$$
$$\leftrightarrow 5 < n.$$

$$\frac{2n}{n+5} < 1.5 \leftrightarrow \frac{4n}{n+5} < 3$$
$$\leftrightarrow 4n < 3n+15$$
$$\leftrightarrow n < 15.$$

Thus all terms b_n meeting both conditions, $n > 5$ and $n < 15$, are in the neighborhood, that is, all terms b_n with $5 < n < 15$. Thus the nine terms $b_6, b_7, b_8, \cdots, b_{14}$ are in the neighborhood $\langle 1, 1.5\rangle$. (See Figure 6.)

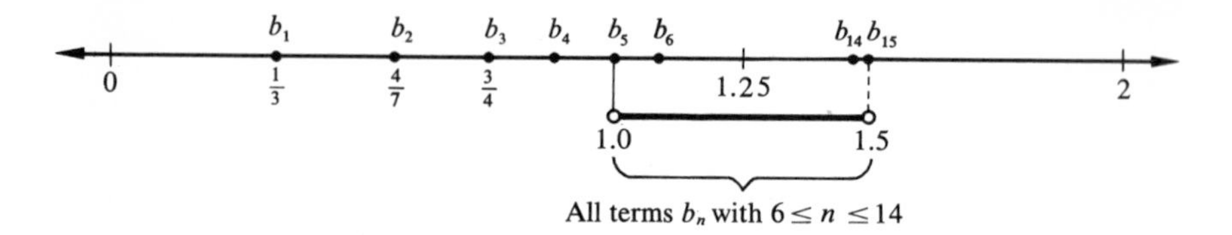

Figure 6. Specifying Which Terms of $\{b_n\} = \left\{\frac{2n}{n+5}\right\}$ *Are in the Neighborhood* $\langle 1, 1.5\rangle$

Example Determine which terms of $\{c_n\} = \left\{\frac{n^2+n}{n^2+10}\right\}$ are in the neighborhood $\langle .9,1\rangle$.

Solution: The inequality $.9 < \frac{n^2+n}{n^2+10} < 1$ must be solved:

$$
\begin{aligned}
.9 < \frac{n^2+n}{n^2+10} &\leftrightarrow 9 < \frac{10n^2+10n}{n^2+10} \\
&\leftrightarrow 9n^2+90 < 10n^2+10n \\
&\leftrightarrow 90 < n^2+10n \\
&\leftrightarrow \frac{90}{n} < n+10 \\
&\leftrightarrow 6 \leq n.
\end{aligned}
$$

$$
\begin{aligned}
\frac{n^2+n}{n^2+10} < 1 &\leftrightarrow n^2+n < n^2+10 \\
&\leftrightarrow n < 10.
\end{aligned}
$$

Thus all terms meeting both conditions, $n \geq 6$ and $n < 10$ — that is, all terms c_n with $6 \leq n < 10$ — are in the neighborhood $\langle .9, 1\rangle$. Thus c_6, c_7, c_8, and c_9 are the terms of $\{c_n\}$ which are contained in the interval $\langle .9, 1\rangle$.

3. Handling Inequalities

In order to gain proficiency in dealing with sequences and neighborhoods, it would be helpful at this point to review certain common *techniques for handling inequalities.* The following examples are designed for this purpose.

It is often helpful to *multiply or divide both members by the same number* in order to clear up decimals or fractions.

Examples (1)

$$\frac{n+1}{2n} < .1 \leftrightarrow \frac{10n+10}{2n} < 1$$
$$\leftrightarrow 10n + 10 < 2n, \text{ etc.}$$

(2)

$$\frac{n+1}{2n} < \frac{2}{3} \leftrightarrow \frac{3n+3}{2n} < 2$$
$$\leftrightarrow 3n + 3 < 4n, \text{ etc.}$$

If, however, we multiply or divide both members by a negative number, then the inequality symbol must be reversed, as in the following.

Examples (3)

$$\frac{n-10}{2n} < -\frac{1}{2} \leftrightarrow (-2)\left(\frac{n-10}{2n}\right) > (-2)\left(-\frac{1}{2}\right)$$
$$\leftrightarrow \frac{20-2n}{2n} > 1$$
$$\leftrightarrow 20 - 2n > 2n, \text{ etc.}$$

(4)

$$-2n^2 > -10 \leftrightarrow (-1)(-2n^2) < (-1)(-10)$$
$$\leftrightarrow 2n^2 < 10, \text{ etc.}$$

(5)

$$\frac{n^2}{1-2n} < 3 \leftrightarrow n^2 > 3(1-2n), \text{ since } 1-2n \text{ is always negative, etc.}$$

If the same number is added to or subtracted from both members, the inequality symbol is not reversed.

Examples (6)

$$2n + 7 < 3n + 4 \leftrightarrow 7 < n + 4$$
$$\leftrightarrow 3 < n.$$

(7)

$$5n - 3 < 6n \leftrightarrow -3 < n.$$

[*Note:* This inequality is true for all natural numbers n, since the left member is negative.]

If a, b, c, and d are positive real numbers, then $\frac{a}{b} < \frac{c}{d} \leftrightarrow \frac{b}{a} > \frac{d}{c}$. Also $\frac{a}{b} < c \leftrightarrow \frac{a}{c} < b$.

Examples (8)

$$\frac{2}{n} < \frac{3}{4} \leftrightarrow \frac{n}{2} > \frac{4}{3}.$$

(9)

$$\frac{1}{n} > 3 \leftrightarrow \frac{n}{1} < \frac{1}{3}.$$

(10) $$\frac{2}{n} < 3 \leftrightarrow \frac{2}{3} < n.$$

(11) $$\frac{7}{n+1} < 11 \leftrightarrow \frac{7}{11} < n + 1.$$

The following examples should also be noted:

Examples (12) $$\sqrt{n+1} > n \leftrightarrow (\sqrt{n+1})^2 > n^2$$
$$\leftrightarrow n + 1 > n^2, \text{ etc.}$$

(13) $$n^2 > 8 \leftrightarrow n > \sqrt{8}$$
$$\leftrightarrow n \geq 3.$$

(14) $$\frac{2n+3}{n+5} < 2 \leftrightarrow 2n + 3 < 2(n + 5)$$
$$\leftrightarrow 2n + 3 < 2n + 10, \text{ etc.}$$

(15) $$\frac{2n+3}{n+5} < 2 + E \leftrightarrow 2n + 3 < (2 + E)(n + 5)$$
$$\leftrightarrow 2n + 3 < 2n + 10 + En + 5E, \text{ etc.}$$

EXERCISES

1. Solve the following inequalities, keeping in mind that n refers to natural numbers only.

(a) $\frac{2n+5}{3n} < 1$

(b) $\frac{2n-11}{n} < 2.1$

(c) $3.4 < \frac{3n+14}{n}$

(d) $-1 < \frac{4-n}{n}$

(e) $\frac{4-n}{n} < -\frac{1}{2}$

(f) $\frac{n-7}{n+3} < .9$

(g) $.95 < \frac{n-7}{n+3}$

(h) $\frac{n^2+4}{3n^2} < \frac{1}{2}$

2. Find which terms of each given sequence are in the neighborhood(s) given.

(a) $\{a_n\} = \left\{\frac{n}{n+5}\right\}$ $\langle .98, 1.02\rangle$

(b) $\{b_n\} = \left\{\frac{1}{n}\right\}$ $\langle -.2, .2\rangle$, $\langle -\frac{3}{20}, \frac{3}{20}\rangle$

(c) $\{c_n\} = \left\{\frac{3n^2}{n^2+1}\right\}$ $\langle 2.8, 3.2\rangle$

4. General Neighborhoods

Sometimes it is desirable to deal with a general neighborhood of a given point — that is, a neighborhood whose radius is not specified. For such a neighborhood the variable E will represent the radius, where *E is any positive real number*. By using E to represent the radius of *any* neighborhood of a point P, the compound inequality $P - E < a_n < P + E$ can be solved for a given sequence $\{a_n\}$, yielding a general solution for n in terms of E.

Example Let $\{a_n\} = \left\{\dfrac{3n}{n+1}\right\}$ and $P = 3$. Solving the compound inequality $3 - E < \dfrac{3n}{n+1} < 3 + E$ can be accomplished as follows:

$$\begin{aligned} 3 - E < \frac{3n}{n+1} &\leftrightarrow (3 - E)(n + 1) < 3n \\ &\leftrightarrow 3n + 3 - En - E < 3n \\ &\leftrightarrow 3 - E < En \\ &\leftrightarrow \frac{3}{E} - 1 < n. \end{aligned}$$

$$\begin{aligned} \frac{3n}{n+1} < 3 + E &\leftrightarrow 3n < (3 + E)(n + 1) \\ &\leftrightarrow 3n < 3n + 3 + En + E \\ &\leftrightarrow -3 - E < En \\ &\leftrightarrow -\frac{3}{E} - 1 < n, \end{aligned}$$

which is true for all natural numbers, since $-\dfrac{3}{E}$ is negative.

Therefore, the inequality $3 - E < \dfrac{3n}{n+1} < 3 + E$ is equivalent to the inequality $n > \dfrac{3}{E} - 1$. For example, taking E to be, successively, .1, .01, and .001, we have $n > \dfrac{3}{.1} - 1 = 29$, $n > \dfrac{3}{.01} - 1 = 299$ and $n > \dfrac{3}{.001} - 1 = 2999$. Thus

$$2.9 < \frac{3n}{n+1} < 3.1 \leftrightarrow n > 29,$$

$$2.99 < \frac{3n}{n+1} < 3.01 \leftrightarrow n > 299,$$

$$2.999 < \frac{3n}{n+1} < 3.001 \leftrightarrow n > 2999, \text{ etc.}$$

In other words, all terms after the 29th term lie in the neighborhood $\langle 2.9, 3.1 \rangle$, all terms after the 299th lie in the neighborhood $\langle 2.99, 3.01 \rangle$, and all terms after the 2999th lie in the neighborhood $\langle 2.999, 3.001 \rangle$. (See Figure 7.) In general, all terms a_n with $n > \frac{3}{E} - 1$ lie in the neighborhood $\langle 3 - E, 3 + E \rangle$.

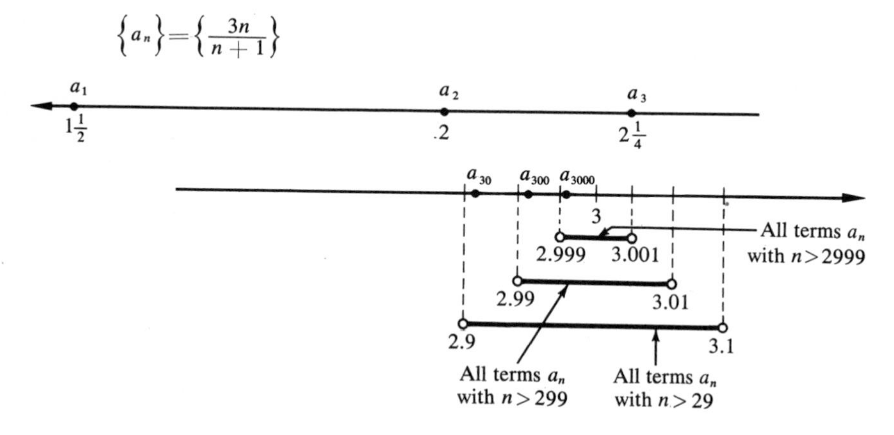

Figure 7

In this example, it was found that $\frac{3n}{n+1} < 3 + E$ is equivalent to $-\frac{3}{E} - 1 < n$. Since *E is always positive*, $-\frac{3}{E} - 1$ is always negative, and certainly every natural number n is greater than any negative number. Thus $-\frac{3}{E} - 1 < n$ is true for all natural numbers, and so is $\frac{3n}{n+1} < 3 + E$. This means, therefore, that all terms of the sequence lie to the *left* of $3 + E$ on the number line.

Moreover, as was shown, $3 - E < \frac{3n}{n+1}$ is equivalent to $\frac{3}{E} - 1 < n$. Thus all terms a_n with $n > \frac{3}{E} - 1$ lie to the *right* of $3 - E$. Consequently, any neighborhood $\langle 3 - E, 3 + E \rangle$ contains all terms with $n > \frac{3}{E} - 1$.

EXERCISES

1. Assume E is any positive real number. Solve each inequality for n.

(a) $\dfrac{2n+3}{3n} < \frac{2}{3} + E$

(b) $\dfrac{7n+11}{n} < 7 + E$

(c) $1 - E < \dfrac{n^2 - 13}{n^2}$

(d) $-E < \dfrac{3}{n-5}$

2. In each of the following, assume E to be any positive real number.

(a) Which terms of $\{a_n\} = \left\{\dfrac{n+3}{2n+3}\right\}$ are less than $\frac{1}{2} + E$? $\frac{1}{2} + \frac{1}{2}$? $\frac{1}{2} + \frac{1}{4}$?

(b) Which terms of $\{b_n\} = \left\{\dfrac{n+1}{2n+3}\right\}$ are less than $\frac{1}{2} + E$? $\frac{1}{2} + \frac{1}{10}$? $\frac{1}{2} + \frac{1}{100}$?

(c) Which terms of $\{d_n\} = \left\{\dfrac{n^2+5}{n^2}\right\}$ are less than $1 + E$? $1 + \frac{1}{2}$? $1 + \frac{1}{3}$? $1 + \frac{1}{4}$?

(d) Which terms of $\{g_n\} = \left\{\dfrac{2n-5}{2n+1}\right\}$ are greater than $1 - E$? $1 - \frac{1}{2}$? $1 - \frac{1}{4}$?

3. Given $\{c_n\} = \left\{\dfrac{3n}{2n-1}\right\}$. Which terms of $\{c_n\}$ are

(a) >1.5?

(b) <1.5?

(c) ≥ 1.6?

(d) in $\langle 1.4, 1.6\rangle$?

(e) in $\langle 1.5 - E, 1.5 + E\rangle$?

(f) in $\langle 1.45, 1.55\rangle$?

4. Given $\{b_n\} = \left\{\dfrac{7n+13}{3n}\right\}$. Which of the terms of $\{b_n\}$ are

(a) >2?

(b) $<2\frac{2}{3}$?

(c) in $\langle 2, 2\frac{2}{3}\rangle$?

(d) $>2\frac{1}{3} - E$?

(e) $<2\frac{1}{3} + E$?

(f) in $\langle 2\frac{1}{3} - E, 2\frac{1}{3} + E\rangle$?

(g) in $\langle \frac{13}{6}, \frac{15}{6}\rangle = \langle 2\frac{1}{3} - \frac{1}{6}, 2\frac{1}{3} + \frac{1}{6}\rangle$?

5. Let $\{a_n\} = \left\{\dfrac{3n-7}{n+1}\right\}$.

(a) Which terms are in the neighborhood $\langle 2.95, 3.05\rangle$?
(b) Which terms are in the neighborhood $\langle 3 - E, 3 + E\rangle$?

5. Limit of a Sequence

Let us first consider the sequence $\{b_n\} = \left\{\frac{2n+1}{n}\right\}$. The first four terms of this sequence are 3, $2\frac{1}{2}$, $2\frac{1}{3}$, and $2\frac{1}{4}$. No matter how many terms of $\{b_n\}$ we actually locate on the real number line, we will never arrive at 2, but we will surely get *closer and closer* to 2. It will make sense, as we shall discuss, to say that *this sequence approaches 2 as a limit*. Certainly, the phrase *closer and closer* is somewhat vague, and mathematical ideas must not be encased in vague language. It is true that the terms are getting closer and closer to 2, but they are also getting closer and closer to 1 and to 0 and indeed to any number less than 2. Why should 2 rather than some other number be the limit?

The answer is that we could not get *as close as we please* to any number other than 2 with terms of the sequence. Let us now determine what is meant by getting *as close as we please* to 2 with terms of the sequence $\{b_n\} = \left\{\frac{2n+1}{n}\right\}$.

Suppose two persons are playing a game in which the following events take place:

The first person chooses any neighborhood of 2 on the number line.

The second person tries to find a particular term of the sequence $\{b_n\} = \left\{\frac{2n+1}{n}\right\}$ which meets two conditions:

(i) This term lies in the neighborhood chosen by the first person,
(ii) All terms of $\{b_n\}$ following this term also lie in the neighborhood.

If the second person can always win — i.e., he can always find such a term — no matter what neighborhood the first person chooses, then it will make sense to say that we can get *as close as we please* to 2 with terms of the sequence.

If the first person should choose the neighborhood $\langle 1.9, 2.1\rangle$, the second person would solve the inequality $1.9 < \frac{2n+1}{n} < 2.1$:

$1.9 < \frac{2n+1}{n} \leftrightarrow 19 < \frac{20n+10}{n}$	$\frac{2n+1}{n} < 2.1 \leftrightarrow \frac{20n+10}{n} < 21$
$\leftrightarrow 19n < 20n + 10$	$\leftrightarrow 20n + 10 < 21n$
$\leftrightarrow -10 < n$ which is true for all natural numbers n.	$\leftrightarrow 10 < n.$

Therefore $1.9 < \frac{2n+1}{n} < 2.1 \leftrightarrow n > 10.$

This means that the second person could choose the 11th term, or any term thereafter, and meet conditions (i) and (ii). You might wonder why the second person could not win by choosing the 10th term, $b_{10} = 2.1$. He could not because our conditions require the second person to find a term such that it is *closer* to 2 than 2.1 is.

If the first person should choose the neighborhood $\langle 1.99, 2.01 \rangle$, the second person would solve the inequality $1.99 < \frac{2n+1}{n} < 2.01$, obtaining the equivalent expression $n > 100$. Thus the second person could choose the 101st term, or any term thereafter, as his choice of a particular term meeting the conditions of the game.

Thus we have shown that for neighborhoods $\langle 1.9, 2.1 \rangle$ and $\langle 1.99, 2.01 \rangle$, the second person could win the game. (See Figure 8.) Now we should try to show that, given *any* neighborhood $\langle 2 - E, 2 + E \rangle$ of 2, the second person could win the game.

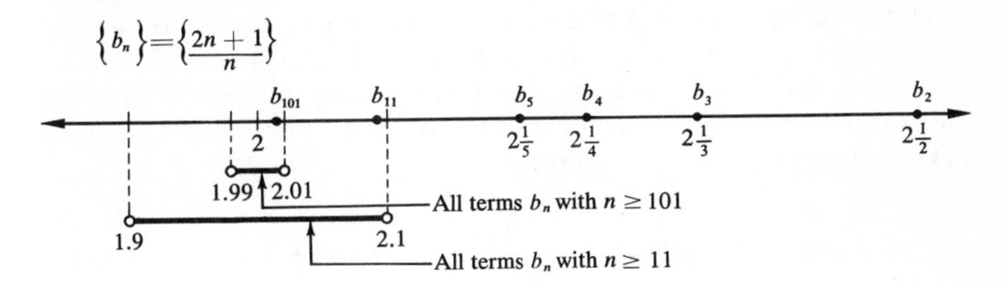

Figure 8

Solving the inequality $2 - E < \frac{2n+1}{n} < 2 + E$ is accomplished as follows:

$$2 - E < \frac{2n+1}{n} \leftrightarrow 2n - En < 2n + 1$$
$$\leftrightarrow -En < 1$$
$$\leftrightarrow n > -\frac{1}{E}.$$

$$\frac{2n+1}{n} < 2 + E \leftrightarrow 2n + 1 < 2n + En$$
$$\leftrightarrow 1 < En$$
$$\leftrightarrow \frac{1}{E} < n.$$

Since E is always positive, then $-\frac{1}{E}$ is always negative and, hence, is less than any natural number. Thus, the inequality $n > -\frac{1}{E}$ is true for all natural numbers, so that $2 - E < \frac{2n+1}{n} < 2 + E \leftrightarrow n > \frac{1}{E}\cdot$ Let us now give an interpretation to this result.

The preceding solution of the inequality $2 - E < \frac{2n+1}{n} < 2 + E$ reveals that the general neighborhood of 2, $\langle 2 - E, 2 + E\rangle$, contains all terms of $\left\{\frac{2n+1}{n}\right\} = \{b_n\}$ with $n > \frac{1}{E}\cdot$ This means that, no matter which neighborhood $\langle 2 - E, 2 + E\rangle$ the first person chooses, the second person could choose any term b_n with $n > \frac{1}{E}$ and it would meet conditions (i) and (ii). For this reason, we say that the terms of $\{b_n\}$ get *as close as we please* to 2, and this is what we mean when we say that the sequence $\{b_n\}$ *approaches* 2 *as a limit.*

REMARK: Using the two neighborhoods $\langle 1.9, 2.1\rangle$ and $\langle 1.99, 2.01\rangle$ was not absolutely necessary in the preceding discussion, since we needed to use only the general neighborhood $\langle 2 - E, 2 + E\rangle$, as in the last paragraph. However, in this and similar situations, your understanding of the ideas involved will probably be clearer if you try one or two particular neighborhoods before working with the general neighborhood.

After obtaining the inequality $n > \frac{1}{E}$ in the preceding example, we might find it enlightening to choose some actual values for E and substitute these values in the inequality $n > \frac{1}{E}$, as suggested in the table.

If $E =$	$\frac{1}{2}$	$\frac{1}{4}$	$\frac{1}{10}$	$\frac{1}{1000}$
Then $\frac{1}{E} =$	2	4	10	1000

Now let us consider another example. The first six terms of the sequence

defined by $c_n = \begin{cases} \frac{n-1}{n} & \text{for } n \text{ odd} \\ \frac{n^2+3}{n^2} & \text{for } n \text{ even} \end{cases}$ are $0, 1\frac{3}{4}, \frac{2}{3}, 1\frac{3}{16}, \frac{4}{5}, 1\frac{1}{12}$.

The terms get *closer and closer* to 1 as n increases, and they approach 1 from *both* the left and the right. (See Figure 9.)

Suppose two persons are playing a game to show that the limit of $\{c_n\}$ is 1. The first person chooses *any* neighborhood of 1 he so desires. Let E be the radius of this neighborhood. The task of the second person is to find a particular term of the sequence such that this term and all succeeding terms are in the neighborhood. In other words, the task of the second person is to find a particular natural number M such that all terms c_n with $n \geq M$ are in the neighborhood. If it is true that no matter what neighborhood the first person chooses the second person can find an appropriate natural number M, then we shall say that the limit of $\{c_n\}$ is 1.

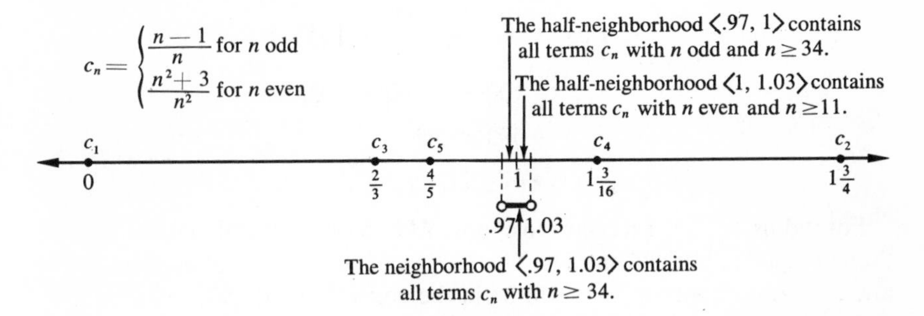

Figure 9

Since there are an infinite number of neighborhoods of 1, it is impossible to test them all. We will test a neighborhood of 1 of radius .03 and then a neighborhood of radius E, where E represents any positive real number. We could forego choosing a particular neighborhood and simply go on to the general case; however, taking at least one particular neighborhood (we often take two) will aid our understanding of the situation, thereby making the general case more meaningful. For radius .03 (see Figure 9) we have for n odd:

$$\begin{aligned} .97 < \frac{n-1}{n} &\leftrightarrow 97 < \frac{100n - 100}{n} \\ &\leftrightarrow 97n < 100n - 100 \\ &\leftrightarrow 100 < 3n \\ &\leftrightarrow n \geq 34, \end{aligned}$$

$$\begin{aligned} \text{and } \frac{n-1}{n} < 1.03 &\leftrightarrow \frac{100n - 100}{n} < 103 \\ &\leftrightarrow 100n - 100 < 103n \\ &\leftrightarrow -100 < 3n \\ &\leftrightarrow -33\tfrac{1}{3} < n, \text{ which is true for} \\ &\qquad \text{all natural numbers } n; \end{aligned}$$

for n even:

$$.97 < \frac{n^2+3}{n^2} \leftrightarrow 97 < \frac{100n^2+300}{n^2}$$
$$\leftrightarrow 97n^2 < 100n^2 + 300$$
$$\leftrightarrow -300 < 3n^2$$
$$\leftrightarrow -100 < n^2, \text{ which is true for all natural numbers } n,$$

$$\text{and } \frac{n^2+3}{n^2} < 1.03 \leftrightarrow \frac{100n^2+300}{n^2} < 103$$
$$\leftrightarrow 100n^2 + 300 < 103n^2$$
$$\leftrightarrow 300 < 3n^2$$
$$\leftrightarrow n \geq 11.$$

For radius .03, it is necessary to choose M to be any natural number greater than or equal to both 34 and 11. That is, M can be any natural number ≥ 34, and all terms c_n with $n \geq M$ will be in the neighborhood $\langle .97, 1.03\rangle$.

For radius E we have for n odd:

$$1 - E < \frac{n-1}{n} \leftrightarrow n - En < n - 1$$
$$\leftrightarrow -En < -1$$
$$\leftrightarrow En > 1$$
$$\leftrightarrow n > \frac{1}{E},$$

$$\text{and } \frac{n-1}{n} < 1 + E \leftrightarrow n - 1 < n + En$$
$$\leftrightarrow -1 < En$$
$$\leftrightarrow -\frac{1}{E} < n, \text{ which is true for all natural numbers } n;$$

for n even:

$$1 - E < \frac{n^2+3}{n^2} \leftrightarrow n^2 - En^2 < n^2 + 3$$
$$\leftrightarrow -En^2 < 3$$
$$\leftrightarrow n^2 > -\frac{3}{E}, \text{ which is true for all natural numbers } n,$$

$$\text{and } \frac{n^2+3}{n^2} < 1 + E \leftrightarrow n^2 + 3 < n^2 + En^2$$
$$\leftrightarrow 3 < En^2$$
$$\leftrightarrow \sqrt{\frac{3}{E}} < n.$$

Let M be any natural number greater than the maximum of the two numbers $\frac{1}{E}$ and $\sqrt{\frac{3}{E}}$. Then all terms c_n with $n \geq M$ will lie in the neighborhood $\langle 1 - E, 1 + E \rangle$. This means that the limit of $\{c_n\}$ is 1.

It is enlightening to choose some particular values of E and determine the corresponding possible values for M, as shown in the following table:

If E is	1	.5	.4	$\frac{1}{3}$	.2	.1	.01
then $\frac{1}{E} =$	1	2	2.5	3	5	10	100
and $\sqrt{\frac{3}{E}} =$	$\sqrt{3}$	$\sqrt{6}$	$\sqrt{7.5}$	3	$\sqrt{15}$	$\sqrt{30}$	$10\sqrt{3}$
Thus choose M to be any natural number >	$\sqrt{3}$	$\sqrt{6}$	$\sqrt{7.5}$	3	5	10	100

EXERCISES

1. Let $\{a_n\} = \left\{\frac{n+1}{2n}\right\}$. Compute enough terms to decide what number appears to be the limit. Find which terms are in the neighborhood of radius .1 of the limit. Find which terms are in the neighborhood of radius .01 of this proposed limit. Do the same for a neighborhood of radius E.

2. (a) Given the sequence $\{b_n\} = \left\{\frac{n-2}{n+2}\right\}$, show that the limit is 1, using neighborhoods $\langle .9, 1.1 \rangle$ and $\langle 1 - E, 1 + E \rangle$.

(b) Using your solution to the inequality $1 - E < \frac{n-2}{n+2} < 1 + E$, determine which terms are in the neighborhoods $\langle 1 - .01, 1 + .01 \rangle$, $\langle 1 - .001, 1 + .001 \rangle$, $\langle 1 - .0001, 1 + .0001 \rangle$.

3. Given $\{c_n\} = \left\{\dfrac{n+3}{n}\right\}$,

(a) Find which terms are in the neighborhood of $\frac{1}{2}$ with radius $\frac{1}{6}$.

(b) Find, if possible, a neighborhood of $\frac{1}{2}$ which does not contain an infinite number of terms of $\{c_n\}$.

(c) What is the limit of $\{c_n\}$? Prove your answer!

4. (a) Show that the limit of $\{d_n\} = \left\{\dfrac{1}{n}\right\}$ is 0, using the neighborhoods $\langle -.03, .03\rangle$ and $\langle -E, E\rangle$.

(b) Determine which terms are in the neighborhoods $\langle -.01, .01\rangle$, $\langle -.001, .001\rangle$ and $\langle -.0005, .0005\rangle$. You need not work out inequalities for these neighborhoods, since you can use the general expression involving E and n which you found for the neighborhood $\langle -E, E\rangle$.

6. Definition of Limit

Let us summarize the concept of limit of a sequence as it has been discussed in the foregoing sections.

Let $\{a_n\}$ be a sequence and A a real number. Suppose two persons are playing a game in which the first person selects any neighborhood of A and the task of the second person is to find a particular natural number M such that a_M and all terms a_n with $n > M$ are in the neighborhood. *If, for every neighborhood of A* (no matter how large or how small) *the first person could choose, the second person can find a particular natural number M for that neighborhood, then the sequence $\{a_n\}$ is said to approach A as a limit*, or, as we shall more commonly say, the sequence $\{a_n\}$ is said to *converge to A*. Thus we are led to formulate the following definitions:

DEFINITION OF CONVERGENCE

A sequence $\{a_n\}$ is said to *converge* to a number A (or to approach A as a limit) if for every neighborhood of A there can be found a natural number M such that all terms of $\{a_n\}$ with $n \geq M$ are in the neighborhood.

RELATED DEFINITIONS

The number A is called the *limit* of the sequence.

If a sequence converges, it is said to be a *convergent sequence*.

The notation $\{a_n\} \to A$ is read, "The sequence $\{a_n\}$ converges to A" or "The sequence $\{a_n\}$ approaches A as a limit."

REMARKS: (1) The capital letter M will be used to refer to a *particular* natural number, as indicated by the definition of convergence. A lower-case n will be used to refer to *any* member of the set of natural numbers or of a subset of the set of natural numbers. For example, if $\{a_n\} = \left\{\frac{1}{n}\right\}$ and $E = .1$, we can choose M to be 11 (or any particular natural number greater than 11), and we can make the statement, "If n is any natural number ≥ 11, then a_n is in the neighborhood $\langle -.1, .1\rangle$." In this case the variable n refers to any member of the set of natural numbers ≥ 11. If no value of E is specified, then we would show that $0 - E < \frac{1}{n} < 0 + E$ is equivalent to $n > \frac{1}{E}$ and we would make the statement, "If M is some particular natural number $> \frac{1}{E}$, then all terms a_n with $n \geq M$ are in the neighborhood $\langle -E, E\rangle$."

(2) For a sequence $\{a_n\}$, a certain number A may appear to be the limit, and the definition of convergence may enable you to prove that A is the limit, provided, of course, the sequence really does have limit A. Determining the *suspected limit* is a matter for which you must use insight, intuition and so-called educated guessing. In subsequent exercises you will have numerous opportunities for investigating sequences and for finding and proving limits.

(3) Students sometimes fail to realize that if $a > b$, then a^2 is not necessarily greater than b^2. If $a = 2$ and $b = -3, -2$, and -1, respectively, we have $2^2 < (-3)^2$, $2^2 = (-2)^2$, and $2^2 > (-1)^2$. The conclusion is that if a is positive and b is negative, then a^2 may be greater than, equal to, or less than b^2. Of course, if a and b are both positive and $a > b$, then it is true that $a^2 > b^2$. (What conclusion can you make in the event a and b are both negative?)

The Transitive Property

If a, b, and c are any three real numbers and $a > b$ and $b > c$, then $a > c$. This property of the real numbers has many applications in the solution of inequalities, as the following examples point out.

Examples **(1)** To solve $\frac{n}{n+3} < 1 + E$, we notice that $\frac{n}{n+3} < 1$ is true for all natural numbers and also that $1 < 1 + E$. So by the Transitive Property, $\frac{n}{n+3} < 1 + E$ is true for all natural numbers.

(2) Consider the inequality $-5 < \frac{2n}{n+5}$. Since $0 < \frac{2n}{n+5}$ is true for all natural numbers and since $-5 < 0$, then by the Transitive Property $-5 < \frac{2n}{n+5}$ is true for all natural numbers.

(3) For the inequality $\frac{5n^2 + 13}{n^2} > 5 - E$, since $\frac{5n^2 + 13}{n^2} = \frac{5n^2}{n^2} + \frac{13}{n^2} = 5 + \frac{13}{n^2} > 5$ for all natural numbers and since $5 > 5 - E$, we have $\frac{5n^2 + 13}{n^2} > 5 - E$ for all natural numbers.

Proving Convergence

The following two examples are presented to suggest ways for writing up subsequent proofs of convergence.

Examples **(1)** To prove that $\{a_n\} = \left\{\frac{n}{n+1}\right\} \to 1$:

$$\begin{aligned} 1 - E < \frac{n}{n+1} &\leftrightarrow (1 - E)(n + 1) < n \\ &\leftrightarrow n + 1 - En - E < n \\ &\leftrightarrow 1 - E < En \\ &\leftrightarrow \frac{1}{E} - 1 < n, \end{aligned}$$

and the inequality $\frac{n}{n+1} < 1 + E$ is true for all natural numbers n by the Transitive Property, since $\frac{n}{n+1} < 1$ for all n and $1 < 1 + E$.

It follows that $1 - E < \frac{n}{n+1} < 1 + E \leftrightarrow n > \frac{1}{E} - 1$.

Therefore, if we let M be any natural number $> \frac{1}{E} - 1$, then all terms a_n with $n \geq M$ will be in the neighborhood $\langle 1 - E, 1 + E\rangle$. This proves that $\{a_n\} \to 1$. It may be helpful to select several values of E:

For $E =$	.1	.01	.001	.0001
$\frac{1}{E} - 1 =$	9	99	999	9999

We also have Figure 10 to help make the meaning of this proof clear.

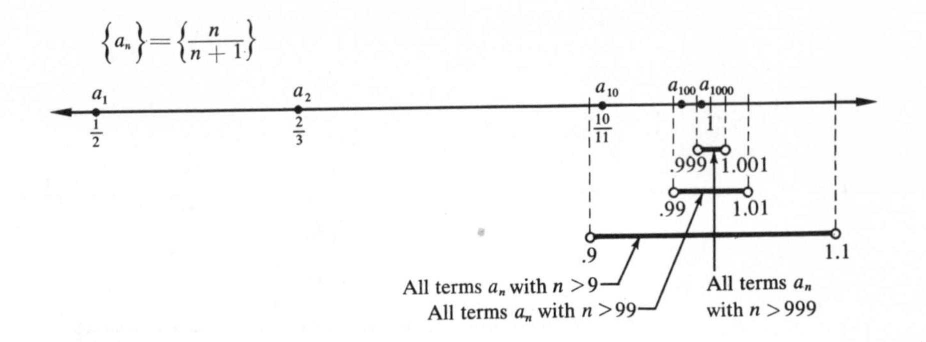

Figure 10

(2) To prove that $\{b_n\} = \left\{\dfrac{2n^2}{n^2+4}\right\} \to 2$:

$\dfrac{2n^2}{n^2+4} < 2 + E \leftrightarrow -\dfrac{8}{E} - 4 < n^2$, which is true for all n.

$$2 - E < \frac{2n^2}{n^2+4} \leftrightarrow (2 - E)(n^2 + 4) < 2n^2$$
$$\leftrightarrow 2n^2 + 8 - En^2 - 4E < 2n^2$$
$$\leftrightarrow \frac{8}{E} - 4 < n^2.$$

If $E \le 2$, $\dfrac{8}{E} - 4$ is nonnegative, so $\dfrac{8}{E} - 4 < n^2 \leftrightarrow \sqrt{\dfrac{8}{E} - 4} < n$.

If $E > 2$, then $\dfrac{8}{E} - 4$ is negative, so that $\dfrac{8}{E} - 4 < n^2$ is true for all natural numbers n.

Therefore, given any neighborhood $\langle 2 - E, 2 + E\rangle$, we can always find an M such that all terms b_n with $n \ge M$ are in this neighborhood: if $0 < E \le 2$, let M be any natural number $> \sqrt{\dfrac{8}{E} - 4}$; and if $E > 2$, let M be any natural number. This proves that $\{b_n\} \to 2$.

It may be helpful to select several values for E:

If $E =$	1	.1	.01	.008
$\sqrt{\dfrac{8}{E} - 4} =$	2	$\sqrt{76}$	$\sqrt{796}$	$\sqrt{996}$

We also have Figure 11 to help make the meaning of this proof clear.

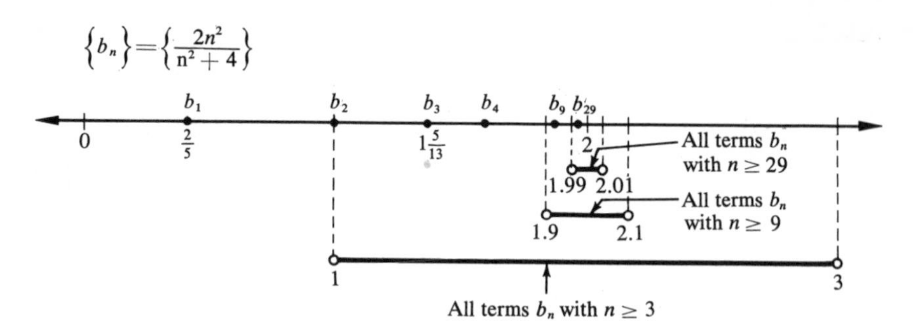

Figure 11

EXERCISES

The first four exercises can be done in a manner suggested by the preceding two examples. Include a graph for each.

1. Prove: $\{a_n\} = \left\{\dfrac{3n}{n+1}\right\} \to 3$.

2. Prove: $\{b_n\} = \left\{\dfrac{3n}{2n-1}\right\} \to \dfrac{3}{2}$.

3. Prove: $\{c_n\} = \left\{\dfrac{n^2+10}{2n^2}\right\} \to \dfrac{1}{2}$.

4. Prove: $\{d_n\} = \left\{\dfrac{2}{\sqrt{n+1}}\right\} \to 0$.

5. Guess the limit of each sequence whose general term is listed. No proof is required.

$$a_n = \frac{7-3n}{2n}; \qquad b_n = \frac{2n^2+3n+1}{5n^2+6n+1}; \qquad c_n = \frac{(n-1)(n-2)}{(n+1)(n+2)};$$

$$d_1 = 8 \text{ and for } n \geq 2,\ d_n = \frac{d_{n-1}}{2};$$

$$f_1 = 8 \text{ and for } n \geq 2,\ f_n = \frac{f_{n-1}}{2} + 1;$$

$$g_n = \frac{\sin n}{n}; \qquad h_n = \frac{\cos n + n}{n}.$$

For each sequence whose general description is given (Exercises 6-10):
(a) Graph the sequence, guess the limit, and find an actual natural number M such that all terms with $n \geq M$ are in the neighborhood of the limit of radius .1.
(b) Then write up a proof of the convergence, as you did in Problems 1–4.

6. $a_n = \dfrac{1}{2n-1}$

7. $b_n = \dfrac{n}{2n-1}$

8. $c_n = \begin{cases} \dfrac{n+1}{n} & \text{for } n \text{ odd} \\ \dfrac{n^2+1}{n^2} & \text{for } n \text{ even} \end{cases}$

9. $d_n = \dfrac{5n+1}{4n+1}$

10. $g_n = \sqrt{\dfrac{4n+1}{n}}$

11. Let $\{q_n\} = \left\{\dfrac{5n+2}{3n}\right\}$.

(a) Which terms of $\{q_n\}$ lie in the neighborhood $\langle 1\frac{1}{3}, 2\rangle$?

(b) Show that for a neighborhood of $1\frac{2}{3}$ with radius E, M can be any natural number $> \dfrac{2}{3E}$.

12. Let $\{a_n\} = \left\{\dfrac{9n+10}{10n}\right\}$.

(a) Show that the neighborhood of 1 with radius .1 contains all terms a_n with $n \geq 6$.

(b) Which terms are in the neighborhood of 1 with radius .04?

(c) To what number, if any, does $\{a_n\}$ converge?

13. Let $\{s_n\} = \left\{\dfrac{3-4n}{n}\right\}$.

(a) Which terms are in the neighborhood $\langle -4.06, -3.94\rangle$?

(b) Solve the inequality $-4 - E < \dfrac{3-4n}{n} < -4 + E$, and use this result to determine which terms are in the neighborhood of -4 of radius .5; .3; .01; and .0001.

7. Some Consequences of the Definition of Convergence

An Alternate Meaning of Convergence

Suppose $\{a_n\}$ is some sequence converging to a number A. According to the definition of convergence, if $\langle A - E, A + E\rangle$ is any neighborhood of A, then there is a particular term a_M such that this term and all succeeding terms are in $\langle A - E, A + E\rangle$. Thus the only terms which can lie *outside* the neighborhood are those among the first $(M - 1)$ terms of the sequence. It is also possible that *no* terms will lie outside the neighborhood. In other words, outside any neighborhood of A there are either no terms or a finite number of terms. More concisely, "If $\{a_n\} \to A$ and $\langle A - E, A + E\rangle$ is *any* neighborhood of A, then at most a finite number of terms of $\{a_n\}$ lie outside $\langle A - E, A + E\rangle$." This logical consequence of the definition of convergence provides an alternate method for determining convergence and will be very useful.

Suppose we have a sequence $\{a_n\}$ and we suspect that it converges to a number A. In order for A to be the limit, it must be true that, for *every* neighborhood of A, all terms following some particular term must be in the neighborhood. This can be stated another way: In order for A to be the limit, it must be true that for *every* neighborhood of A at most a finite number of terms of the sequence lie outside this neighborhood. If we could find *just one* neighborhood of A outside of which there are an infinite number of terms of the sequence, then the sequence would not converge to A.

There is a point of logic involved here which can be illustrated by an example from life. The statement "Every student in this school attended the football game last night" is false if a single student did not attend the game. Likewise, to prove that a given sequence does not converge to a given number, it suffices to find *one* neighborhood of the given number such that an infinite number of terms of the sequence lie outside this neighborhood.

Sequences Having Terms Equal to the Limit

A careful examination of the definition of convergence reveals that a convergent sequence can have one or more of its terms equal to the limit. For example:

(1) Let $a_n = \begin{cases} 3 \text{ for } n \text{ odd} \\ \dfrac{3n+1}{n} \text{ for } n \text{ even.} \end{cases}$

The limit is 3. Every neighborhood of 3 contains the limit 3, of course, and hence all odd terms, a_1, a_3, a_5, etc. Also, every neighborhood $\langle 3 - E, 3 + E\rangle$ contains all even terms a_n with $n > \dfrac{1}{E}\cdot$

(2) Any constant sequence $\{a_n\} = \{c\}$ converges to c because, for any neighborhood $\langle c - E, c + E\rangle$, M (in the definition of convergence) can be the number 1. In other words, outside of any neighborhood of c there are no terms of the sequence, and hence at most a finite number of terms. (See the preceding discussion, An Alternate Meaning of Convergence.)

More About M

It is important to understand the remark made earlier that the capital letter M is used, in discussing convergence, to indicate a particular natural number, whereas a lower-case n is used to refer to any member of the set of natural numbers or of a proper subset of the set of natural numbers. (See Remark (1), page 34.)

Furthermore, you should not receive the impression that for a given sequence and a given neighborhood there is only one natural number M, or that M must be the *first* such natural number meeting the condition of the definition of convergence. (We recall that a sequence $\{a_n\}$ is said to converge to a number A if, for every neighborhood of A, there can be found a natural number M such that all terms of $\{a_n\}$ with $n \geq M$ are in the neighborhood.) Often it is inconvenient or extremely difficult to find the first (that is, the smallest) such natural number M. The definition of convergence requires only that we be able to find some particular natural number M, not necessarily the smallest one. The following examples illustrate this.

Examples **(1)** Given: The sequence $\{a_n\} = \left\{\dfrac{3}{\left(2n + \dfrac{17}{n}\right)^2}\right\}$ and the neighborhood $\langle -.005, .005\rangle$.

Then $-.005 < \dfrac{3}{\left(2n + \dfrac{17}{n}\right)^2} < .005 \leftrightarrow \dfrac{3}{\left(2n + \dfrac{17}{n}\right)^2} < .005$

$$\leftrightarrow \frac{3}{.005} < \left(2n + \frac{17}{n}\right)^2$$

$$\leftrightarrow 600 < \left(2n + \frac{17}{n}\right)^2$$

For all natural numbers n, $2n + \dfrac{17}{n} > 2n$, and therefore

$$\left(2n + \frac{17}{n}\right)^2 > (2n)^2,$$

so that whenever $(2n)^2 > 600$ is true, then also $\left(2n + \dfrac{17}{n}\right)^2 > 600$ is true. Thus it suffices to solve $(2n)^2 > 600$, obtaining

$$4n^2 > 600 \leftrightarrow n^2 > 150$$
$$\leftrightarrow n \geq 13.$$

Therefore, M can be 13 (or any natural number > 13), and all terms a_n with $n \geq M$ are in the neighborhood $\langle -.005, .005\rangle$.

(2) Given: The sequence $\{b_n\} = \left\{\dfrac{1}{n(2n + 13)}\right\}$ and the neighborhood $\langle -.01, .01\rangle$.

$$\text{Then} \quad -.01 < \frac{1}{n(2n+13)} < .01 \leftrightarrow \frac{1}{n(2n+13)} < .01$$
$$\leftrightarrow n(2n+13) > 100.$$

Now, whenever $n > 100$, also $n(2n + 13) > 100$, so that M can be 101 (or any natural number > 101). We ignored the expression $2n + 13$ in the inequality $n(2n + 13)$ temporarily, since this expression always has a value greater than 1. In general, in searching for M, we can ignore any factor to the left of the $>$ symbol which is always greater than or equal to 1. In this problem we could have ignored the expression n in the inequality $n(2n + 13) > 100$, obtaining $2n + 13 > 100$, which is equivalent to $n > \frac{87}{2}$. Thus any $n \geq 44$ will do.

(3) Given: The sequence $\{c_n\} = \left\{\frac{n^2}{n^2+n+3}\right\}$ and the neighborhood $\langle .9, 1.1\rangle$.

$$\text{Then} \quad .9 < \frac{n^2}{n^2+n+3} < 1.1 \leftrightarrow .9 < \frac{n^2}{n^2+n+3}$$
$$\leftrightarrow 9n^2 + 9n + 27 < 10n^2$$
$$\leftrightarrow 27 < n^2 - 9n.$$

Suppose we factor $n^2 - 9n$ into $n(n - 9)$, obtaining $n(n - 9) > 27$. We now have two factors, n and $n - 9$, whose product must be > 27. There are several slightly different ways we can proceed:

(a) We could insist that each factor be > 27, for certainly then their product would be > 27. In this event, n must be > 27 *and* $n - 9 > 27$. That is, $n > 27$ and $n > 36$. Thus take M to be any natural number > 36.

(b) On the other hand, we could insist that one factor be > 1 and the other factor > 27, for this would make their product > 27. Suppose we insist that n be > 27 *and* $n - 9 > 1$. That is, $n > 27$ and $n > 10$. Then we can take M to be any natural number > 27. Or suppose we insist that n be > 1 and $n - 9 > 27$. That is, $n > 1$ and $n > 36$. This means that M can be any natural number > 36.

(c) There is at least one additional way of finding M for the inequality $n^2 - 9n > 27$. We can write

$$n^2 - 9n > 27 \leftrightarrow n^2 > 9n + 27$$
$$\leftrightarrow n > 9 + \frac{27}{n},$$

and for $n > 27$, $\frac{27}{n}$ is less than 1, and hence $9 + \frac{27}{n} < 10$. Thus the inequality is true for $n > 27$, so choose M to be any natural number > 27.

(4) Given: The sequence $\{d_n\} = \left\{\frac{\sqrt{n} + 2}{n}\right\}$ and the neighborhood $\langle -.1, .1\rangle$.

Then
$$-.1 < \frac{\sqrt{n}+2}{n} < .1 \leftrightarrow \frac{\sqrt{n}+2}{n} < .1$$
$$\leftrightarrow 10\sqrt{n} + 20 < n.$$

We can divide both members by $\sqrt{n}$, obtaining $10 + \frac{20}{\sqrt{n}} < \sqrt{n}$.

Whenever $n > 400$, $\sqrt{n} > 20$ and therefore $\frac{20}{\sqrt{n}} < 1$, so that $10 + \frac{20}{\sqrt{n}} < 11 < \sqrt{n}$. Thus choose M to be 401, or any larger natural number.

(5) Given: The sequence $\{g_n\} = \left\{\frac{7}{n^2 + n}\right\}$ and the neighborhood $\langle -.1, .1\rangle$.

Then
$$-.1 < \frac{7}{n^2+n} < .1 \leftrightarrow \frac{7}{n^2+n} < .1$$
$$\leftrightarrow \frac{7}{.1} < n^2 + n$$
$$\leftrightarrow 70 < n^2 + n.$$

Whenever $n > 70$, then $n^2 + n > 70$, so that we can choose $M = 71$ (or any larger natural number). Or, we could ignore the n and, realizing that whenever $n^2 > 70$ then $n^2 + n > 70$, choose $M = 9$ (or any larger natural number). If the sequence is $\left\{\frac{7}{n^2 - n}\right\}$ instead, and the resulting inequality is $n^2 - n > 70$, we could not ignore the n, since it is not true that $n^2 - n$ is always greater than n^2. (In the preceding example, $n^2 + n$ is always greater than n^2, and for this reason we were able to ignore the n.) We can, however, factor $n^2 - n$, obtaining $n(n - 1) > 70$; and whenever $n - 1 > 70$, then $n(n - 1) > 70$. Thus we can choose M to be 72 (or any larger natural number).

EXERCISES

For each sequence whose general term is given and for each neighborhood given, find a natural number M such that all terms of the sequence with $n \geq M$ are in the neighborhood.

1. $a_n = \dfrac{5}{(n+3)^2}$ and $\langle -.01, .01 \rangle$
2. $b_n = \dfrac{1}{n^2+n}$ and $\langle -.01, .01 \rangle$
3. $c_n = \dfrac{2}{n^2+2^n}$ and $\langle -.04, .04 \rangle$
4. $d_n = \dfrac{2n}{n^3+5}$ and $\langle -.02, .02 \rangle$
5. $f_n = \dfrac{n+3}{(\sqrt{n}+2)^2}$ and $\langle .9, 1.1 \rangle$
6. $g_n = \dfrac{\sqrt{n}}{n+1}$ and $\langle -.03, .03 \rangle$
7. $h_n = \dfrac{n-1}{n^2}$ and $\langle -.1, .1 \rangle$

8. Divergent Sequences

DEFINITION

If a sequence does not converge, it is said to *diverge* and is called a *divergent sequence.*

Showing Divergence by Comparison with $\{a_n\} = \{n\}$ and $\{s_n\} = \{-n\}$

Let us assume (1) the sequence $\{a_n\} = \{n\}$ diverges, and (2) if $\{b_n\}$ is a sequence such that $b_n \geq a_n$ for all natural numbers n greater than or equal to some particular natural number M, then $\{b_n\}$ also diverges. (We will not prove that the sequence $\{a_n\} = \{n\}$ diverges; however, using the definition of convergence given, you should be able to construct an argument for its divergence.)

For example, let $\{b_n\} = \{2n\}$. Then for all natural numbers we have $b_n > a_n$, and hence $\{b_n\} = \{2n\}$ diverges by comparison with $\{a_n\} = \{n\}$. Similarly, let $\{c_n\} = \{2n - 3\}$. Then

$$\begin{aligned} 2n - 3 \geq n &\leftrightarrow 2n \geq n + 3 \\ &\leftrightarrow n \geq 3, \end{aligned}$$

which is true for all natural numbers ≥ 3. Thus, beginning with $n = 3$, $c_n \geq a_n$, so $\{c_n\} = \{2n - 3\}$ diverges by comparison with $\{a_n\} = \{n\}$.

Likewise, $\{d_n\} = \{n^2 - 10\}$ diverges, since $n^2 - 10 \geq n$ for all natural numbers ≥ 4. To show that this is true, it may be helpful to write $n^2 - 10 \geq n$

as $n - \frac{10}{n} \geq 1$ or as $n \geq \frac{10}{n} + 1$. Now it can easily be observed that the inequality is true for $n \geq 10$, at least. In this case closer inspection is possible (though not necessary) and reveals that the inequality is true for $n \geq 4$.

We can also assume that $\{s_n\} = \{-n\}$ diverges and that if $\{t_n\}$ is a sequence such that, for all natural numbers n greater than or equal to some particular natural number M, $t_n \leq s_n$, then $\{t_n\}$ diverges.

For example, $12 - 2n^2 \leq -n \leftrightarrow 12 + n \leq 2n^2$

$$\leftrightarrow \frac{12}{n} + 1 \leq 2n, \text{ which is true for } n \geq 3.$$

Thus $\{t_n\} = \{12 - 2n^2\}$ diverges by comparison with $\{s_n\} = \{-n\}$.

REMARK: All these sequences either increase without bound, and hence diverge, or decrease without bound, and hence diverge. In the next chapter, we shall discuss in detail the concept of increasing (and decreasing) sequences. At that time, we will specify precisely what is meant by the phrases *increasing without bound* and *decreasing without bound.*

A Second Method for Showing Divergence

In order for a sequence to diverge, it is not essential that its terms steadily increase or decrease. Consider the sequence $\{c_n\} = \left\{\frac{1}{n} + (-1)^n\right\}$, whose first six terms are 0, $1\frac{1}{2}$, $-\frac{2}{3}$, $1\frac{1}{4}$, $-\frac{4}{5}$, and $1\frac{1}{6}$. All the terms are greater than -1 and no greater than $1\frac{1}{2}$. Let us show why this sequence fails to have a limit. To do so, we need to recall the discussion at the beginning of Section 7 of this chapter, in which it was realized that, "If $\{a_n\} \to A$ and $\langle A - E, A + E\rangle$ is *any* neighborhood of A, then at most a finite number of terms of $\{a_n\}$ lie outside $\langle A - E, A + E\rangle$."

We can easily show by the following method that neither -1 nor 1 is the limit of $\{c_n\}$: Choose a neighborhood of -1 and a neighborhood of 1, each with radius less than one-half the distance from -1 to 1 (that is, less than $\frac{1}{2}|(-1) - (1)|$). Suppose the neighborhoods are $\langle -1.1, -.9\rangle$ and $\langle .9, 1.1\rangle$. Each of these neighborhoods contains an infinite number of terms of the sequence. Thus we have found a neighborhood of -1 outside of which there are an infinite number of terms, namely the terms in $\langle .9, 1.1\rangle$; this means that -1 is not the limit of the sequence. Similarly, we have found a neighborhood of 1 outside of which there are an infinite number of terms, namely the terms in $\langle -1.1, -.9\rangle$; this means that 1 is not the limit. Thus we have shown that neither -1 nor 1 is the limit of the sequence.

Now let L be any number besides -1 and 1. We can choose a neighborhood of L which has no points in common with one, or possibly both, of the neighborhoods $\langle -1.1, -.9\rangle$ and $\langle .9, 1.1\rangle$. Then outside the chosen neighborhood of L there are an infinite number of terms of the sequence, namely those in one or both of the neighborhoods $\langle -1.1, -.9\rangle$ and $\langle .9, 1.1\rangle$. Thus L is not the limit, and we conclude that the sequence has no limit.

This discussion leads to the following theorem.

THEOREM 2-1

Let $\{p_n\}$ be a sequence and A and B two different real numbers. Let $\langle a, b\rangle$ and $\langle c, d\rangle$ be neighborhoods of A and B respectively, with radii less than $\frac{1}{2}|A - B|$. If each of these neighborhoods contains an infinite number of terms of $\{p_n\}$, then $\{p_n\}$ diverges.

Proof

The neighborhoods $\langle a, b\rangle$ and $\langle c, d\rangle$ have no points in common. Outside $\langle a, b\rangle$ there are an infinite number of terms, namely those terms in $\langle c, d\rangle$. This means that A is not the limit. Outside $\langle c, d\rangle$ there are an infinite number of terms, namely those terms in $\langle a, b\rangle$. This means that B is not the limit.

Let L be any number other than A or B. Choose a neighborhood of L such that this neighborhood has no points in common with one, or possibly both, of the neighborhoods $\langle a, b\rangle$ and $\langle c, d\rangle$. Then outside this neighborhood of L there are an infinite number of terms of $\{p_n\}$, namely those terms in either $\langle a, b\rangle$ or $\langle c, d\rangle$. This means that L cannot be the limit.

Thus $\{p_n\}$ has no limit.

Applications of Theorem 2-1

(1) The first six terms of $\{b_n\}$, where $b_n = \begin{cases} 2 + \dfrac{1}{n} & \text{for } n \text{ odd} \\ 3 - \dfrac{2}{n} & \text{for } n \text{ even} \end{cases}$

are 3, 2, $2\frac{1}{3}$, $2\frac{1}{2}$, $2\frac{1}{5}$, $2\frac{2}{3}$.

It seems that 2 and 3 are likely candidates for the limit. Choose a neighborhood of 2 and a neighborhood of 3 so that these neighborhoods have no points in common — say $\langle 1.6, 2.4\rangle$ and $\langle 2.7, 3.3\rangle$. Inside $\langle 1.6, 2.4\rangle$ lie b_3, b_5, b_7, and so on — that is, all terms b_n with n an odd natural number ≥ 3. Thus an infinite number of terms are inside $\langle 1.6, 2.4\rangle$. Inside $\langle 2.7, 3.3\rangle$ lie b_1 and all terms b_n with n an even natural number ≥ 8. Thus an infinite number of terms are inside $\langle 2.7, 3.3\rangle$. Therefore, by Theorem 2-1, $\{b_n\}$ diverges. (See Figure 12.)

(2) Let $\{c_n\} = \{(-1)^n\}$, a sequence whose terms are alternately -1 and 1. It is essential to bear in mind that the terms of any sequence form an *ordered*,

infinite set of real numbers. The fact that only two numbers, -1 and 1, are needed to represent the terms does not mean that the number of terms in the sequence is finite. Even a constant sequence has an infinite number of (identical) terms. It is also important to realize that the sequence $\{(-1)^{n+1}\}$ is a different sequence from $\{(-1)^n\}$.

Let us show why $\{c_n\} = \{(-1)^n\}$ diverges. The likely candidates for the limit are -1 and 1. Choose a neighborhood of -1 and a neighborhood of 1 so that these two neighborhoods have no points in common – say, $\langle -1.5, -.5\rangle$ and $\langle .5, 1.5\rangle$. The neighborhood $\langle -1.5, -.5\rangle$ contains all odd terms of $\{c_n\}$, and hence an infinite number of terms. The neighborhood $\langle .5, 1.5\rangle$ contains all even terms of $\{c_n\}$, and hence an infinite number of terms. Consequently, by Theorem 2-1, $\{c_n\}$ diverges.

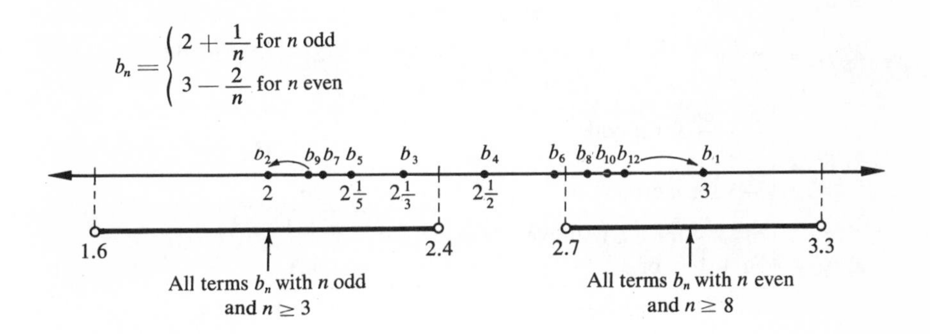

Figure 12

Use of Indirect Proof in Showing Divergence. (*Optional*)

It is easy to realize intuitively that the sequence $\{\sqrt{n}\}$ diverges. An indirect proof of the divergence of this sequence is possible, however, and may be of interest to you.

Proof That $\{a_n\} = \{\sqrt{n}\}$ *Diverges:*

Suppose on the contrary that $\{a_n\}$ converges to some number A. Choose the neighborhood $\langle A - .1, A + .1\rangle$. By definition of convergence there is a particular natural number M such that a_M and all succeeding terms are in this neighborhood. Let J be some natural number $> M$ which is a perfect square, and let K be the first natural number $> J$ which is also a perfect square. Then $a_J = \sqrt{J}$ is a natural number and $a_K = \sqrt{K} = \sqrt{J} + 1 = a_J + 1$. Since a_J and a_K are both in $\langle A - .1, A + .1\rangle$, their difference $a_K - a_J$ must be less than .2. This is impossible, since it was shown above that $a_K = a_J + 1$, and hence $a_K - a_J = 1$. Thus we have reached a contradiction, and our assumption that $\{a_n\}$ converges is false. Therefore, $\{a_n\}$ diverges.

EXERCISES

Investigate for convergence or divergence each of the sequences whose description is given.

If the sequence converges, follow the procedure used in Exercises 1–4, page 38.

If the sequence diverges, show whether the divergence is known by comparison with $\{n\}$ or $\{-n\}$, by Theorem 2-1, by use of indirect proof, or by some other method you might discover.

1. $a_n = \begin{cases} 1 \text{ for } n \text{ odd} \\ \dfrac{2}{n} \text{ for } n \text{ even} \end{cases}$

2. $c_n = 2^n - 3n$

3. $d_n = n^2 - n$

4. $f_n = \sin \dfrac{n}{2}\pi + \dfrac{(-1)^n}{n}$

5. $g_n = \begin{cases} \dfrac{5 - 3n}{n} \text{ for } n \text{ odd} \\ -3 \text{ for } n \text{ even} \end{cases}$

6. $h_n = \begin{cases} \dfrac{4n + 3}{n + 1} & \text{for } n \text{ a multiple of 3} \\ 5 & \text{otherwise} \end{cases}$

7. $j_n = n^{n-5}$

8. $k_n = \dfrac{1}{2^n}$ [Hint: Use logarithms.]

9. $p_n = 4 + \dfrac{(-1)^n}{3}$

10. $q_n = \begin{cases} 2 - \dfrac{1}{n} \text{ for } n \text{ odd} \\ 2 + \dfrac{1}{n^3} \text{ for } n \text{ even} \end{cases}$

11. $r_n = \dfrac{2n}{n + 1}$

12. $s_n = \dfrac{2n^2}{n^2 + 1}$

3 • *Increasing and Decreasing Sequences*

1. Increasing Sequences

DEFINITION

A sequence $\{a_n\}$ is said to be *everywhere increasing* if no term is greater than the succeeding term — that is, for every natural number n, $a_n \leq a_{n+1}$.

Examples *Sequences which are everywhere increasing*

Sequence	*First Six Terms*
$\{a_n\} = \{2n - 1\}$	1, 3, 5, 7, 9, 11
$\{b_n\} = \{\sqrt{n}\}$	1.000, 1.414, 1.732, 2.000, 2.236, 2.449 (approximated to 3 decimal places)
$\{c_n\} = \left\{n + \frac{1}{n}\right\}$	$2, 2\frac{1}{2}, 3\frac{1}{3}, 4\frac{1}{4}, 5\frac{1}{5}, 6\frac{1}{6}$
$\{d_n\} = \{2n + (-1)^{n+1}\}$	3, 3, 7, 7, 11, 11
$\{e_n\} = \left\{\frac{2n}{n+1}\right\}$	$1, 1\frac{1}{3}, 1\frac{1}{2}, 1\frac{3}{5}, 1\frac{2}{3}, 1\frac{5}{7}$
$\{f_n\} = \left\{-\frac{3}{n}\right\}$	$-3, -1\frac{1}{2}, -1, -\frac{3}{4}, -\frac{3}{5}, -\frac{1}{2}$
$\{g_n\} = \left\{3 - \frac{1}{n}\right\}$	$2, 2\frac{1}{2}, 2\frac{2}{3}, 2\frac{3}{4}, 2\frac{4}{5}, 2\frac{5}{6}$

Sometimes, computing the first few terms of a sequence may lead us to suspect that the sequence is everywhere increasing. Certainly, however, there should be at our disposal procedures for actually proving whether or not a given sequence is everywhere increasing.

If the sequence can be described by a single general term, then usually the simplest procedure is to form the inequality $a_n \leq a_{n+1}$, where a_n is the general term, and determine for what natural numbers this inequality is true. If the inequality is true for all natural numbers n, then the sequence is everywhere increasing.

Other procedures are suggested in the examples which follow.

Examples **(1)** If $a_n = \dfrac{2n}{n+3}$, then

$$\begin{aligned}
a_n \leq a_{n+1} &\leftrightarrow \frac{2n}{n+3} \leq \frac{2(n+1)}{(n+1)+3} \\
&\leftrightarrow \frac{2n}{n+3} \leq \frac{2n+2}{n+4} \\
&\leftrightarrow \frac{2n(n+4)}{(n+3)(n+4)} \leq \frac{(2n+2)(n+3)}{(n+4)(n+3)} \\
&\leftrightarrow 2n(n+4) \leq (2n+2)(n+3) \\
&\leftrightarrow 2n^2 + 8n \leq 2n^2 + 8n + 6 \\
&\leftrightarrow 0 \leq 6, \text{ which is always true, and therefore true for all natural numbers } n.
\end{aligned}$$

Thus the sequence is everywhere increasing.

(2) If $b_n = n - \dfrac{1}{n}$, then

$$\begin{aligned}
b_n \leq b_{n+1} &\leftrightarrow n - \frac{1}{n} \leq (n+1) - \frac{1}{n+1} \\
&\leftrightarrow -\frac{1}{n} \leq 1 - \frac{1}{n+1} \\
&\leftrightarrow \frac{-(n+1)}{n(n+1)} \leq \frac{n(n+1) - n}{n(n+1)} \\
&\leftrightarrow -(n+1) \leq n(n+1) - n \\
&\leftrightarrow 0 \leq n^2 + n + 1, \text{ which is true for all natural numbers } n.
\end{aligned}$$

Thus $\{b_n\} = \left\{n - \dfrac{1}{n}\right\}$ is everywhere increasing.

(3) If $c_n = 2^n$, then

$$c_n \leq c_{n+1} \leftrightarrow 2^n \leq 2^{n+1}$$
$$\leftrightarrow 1 \leq 2, \text{ by division by } 2^n, \text{ since } 2^{n+1} = 2^n \cdot 2.$$

Thus $\{2^n\}$ is everywhere increasing.

(4) If $\{d_n\} = \left\{\sin \dfrac{n}{n+1}\right\}$, then we can show that $d_n \leq d_{n+1}$ for all n by the following argument. From our past knowledge of trigonometry, we know that if x_1 and x_2 are any two real numbers in the interval $[\frac{1}{2}, 1]$, with $x_1 \leq x_2$, then $\sin x_1 \leq \sin x_2$.

Thus if we can show

(a) for all natural numbers n, $\dfrac{n}{n+1}$ is in $[\frac{1}{2}, 1]$ and

(b) $\dfrac{n}{n+1} \leq \dfrac{(n+1)}{(n+1)+1}$ for all values of n,

then we will have established that $\sin \dfrac{n}{n+1} \leq \sin \dfrac{(n+1)}{(n+1)+1}$ for all natural numbers n.

(a)

$$\frac{1}{2} \leq \frac{n}{n+1} \leftrightarrow n + 1 \leq 2n$$
$$\leftrightarrow 1 \leq n, \text{ which is true for all natural numbers } n;$$

$$\frac{n}{n+1} \leq 1 \leftrightarrow n \leq n + 1$$
$$\leftrightarrow 0 \leq 1, \text{ which is always true.}$$

Thus $\dfrac{n}{n+1}$ is in the interval $[\frac{1}{2}, 1]$ for all values of n.

(b)

$$\frac{n}{n+1} \leq \frac{(n+1)}{(n+1)+1} \leftrightarrow \frac{n}{n+1} \leq \frac{n+1}{n+2}$$
$$\leftrightarrow n^2 + 2n \leq n^2 + 2n + 1$$
$$\leftrightarrow 0 \leq 1, \text{ which is always true.}$$

Thus, $\dfrac{n}{n+1} \leq \dfrac{(n+1)}{(n+1)+1}$ for all values of n.

(5) Any constant sequence, such as $\{g_n\} = \{5\}$, is everywhere increasing, since each term is no less than the preceding term.

EXERCISES

Determine whether or not each sequence is everywhere increasing. Justify your answer. [Note: Keep the results of this exercise for Exercise 2 of the next section.]

1. $a_n = \sqrt{\dfrac{1}{n}}$
2. $b_n = \dfrac{n+5}{3n}$
3. $c_n = 2n + (-1)^n$
4. $d_n = \sqrt{4 + \dfrac{1}{n}}$
5. $e_n = \dfrac{n^2}{n+1} - n$
6. $g_n = 3^n - 2^n$
7. $h_n = \cos\left(-\dfrac{1}{n}\right)$
8. $j_n = 3n + \cos\dfrac{1}{n}$

2. Sequences Which Increase Without Bound

In Chapter 2 we discussed certain sequences such as $\{a_n\} = \{n\}$, $\{b_n\} = \{n^2 - 10\}$ and $\{c_n\} = \{2n\}$, which were said to *increase without bound.* Let us specify precisely what is meant by this phrase. To do so, we first need the following definitions:

DEFINITIONS

A sequence $\{a_n\}$ is said to be *bounded above* if there is a real number B such that no term of the sequence is greater than B. That is, for all natural numbers n, $a_n \leq B$. Such a number B is called an *upper bound* for the sequence.

If a sequence $\{a_n\}$ is not bounded above, then $\{a_n\}$ is said to be *unbounded above.*

REMARKS: (1) It follows that if a number B is an upper bound for a given sequence, then any number greater than B is also an upper bound for that sequence. Thus any sequence which is bounded above has an infinite number of upper bounds.

(2) The preceding definitions apply to the terms of sequences in general, not merely to the terms of increasing sequences. In fact, these definitions could be stated for any subset of the set of real numbers, and not merely for the set of terms of a sequence. However, since we are not concerned here with sets in general, we have limited our definitions to sequences.

Examples The sequence $\{a_n\} = \{-n\}$ is bounded above by -1 and also by any number greater than -1. The sequences $\{b_n\} = \left\{\frac{1}{n}\right\}$, $\{c_n\} = \left\{\frac{n}{n+1}\right\}$ and $\{d_n\} = \{(-1)^n\}$ are each bounded above by 1 and also by any number greater than 1. (See Figure 13.)

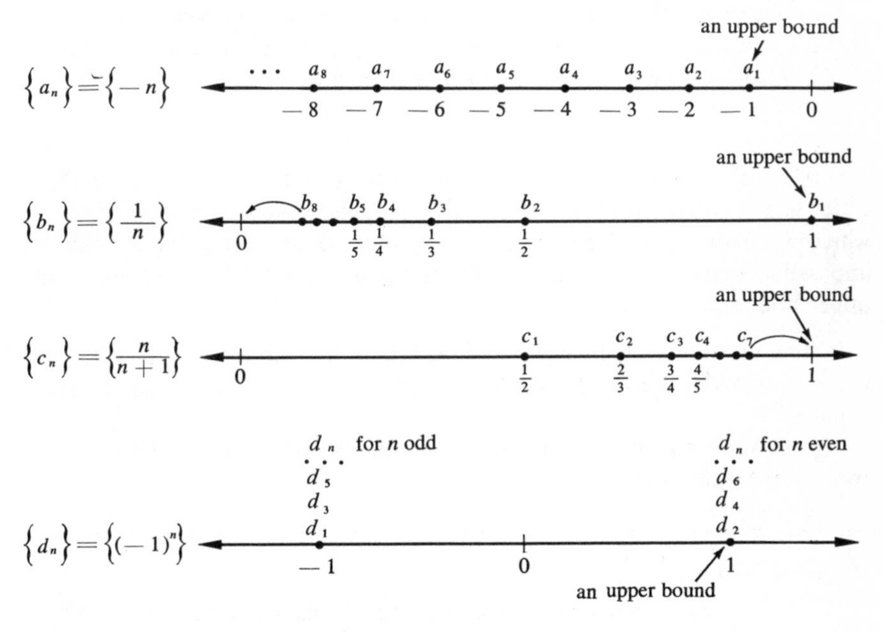

Figure 13. Examples of Sequences Which Are Bounded Above

Now that we have specified what it means for a sequence to be *everywhere increasing*, and also what it means for a sequence to be *unbounded above*, we can formulate the following definition:

DEFINITION

If a sequence is everywhere increasing and is unbounded above, then the sequence is said to *increase without bound*.

REMARKS: (1) As was discussed in the preceding chapter, one way to show that a sequence diverges is to show that the sequence increases without bound.

(2) The proof given for the next theorem is an *indirect proof*. We ask ourselves the question, "Is it possible for this theorem to be false?" and find

that it is *not* possible, since some impossible situation would occur. Indirect proofs are quite common in mathematics. In high school they are particularly common in courses in plane geometry.

THEOREM 3-1

Comparison Principle. Let $\{a_n\}$ be a sequence which increases without bound and let $\{b_n\}$ be a sequence which is everywhere increasing. If there is a particular natural number M such that, for all natural numbers $n \geq M$, $b_n \geq a_n$, then $\{b_n\}$ also increases without bound.

Proof

Suppose the theorem were false. Then there would exist a sequence $\{b_n\}$ which is everywhere increasing and which is bounded above by some number which we shall call P. Then P is also an upper bound for $\{a_n\}$; but this is impossible since $\{a_n\}$ increases without bound and thus cannot have an upper bound.

In the preceding chapter we assumed that the sequence $\{n\}$ increases without bound. We shall use this assumption extensively in the present chapter.

The notion of a subsequence is important to our present discussion, as well as to future discussions.

DEFINITION

Let $\{a_n\}$ be a sequence. A sequence $\{b_n\}$ is called a *subsequence* of $\{a_n\}$ if the following conditions are met:

(i) Every term of $\{b_n\}$ is also a term of $\{a_n\}$, and (ii) The same order is preserved. That is, if b_p and b_q are any two distinct terms of $\{b_n\}$ with $p < q$ and $b_p = a_r$ and $b_q = a_s$, then we must have $r < s$.

For example, $\{b_n\} = \{n\}$ is a subsequence of $\{a_n\} = \{\sqrt{n}\}$; $b_1 = a_1 = 1$, $b_2 = a_4 = 2$, $b_3 = a_9 = 3$, etc. Also, if $\{a_n\} = \left\{\dfrac{1}{n+5}\right\}$ and $\{b_n\} = \left\{\dfrac{1}{n+15}\right\}$, then $b_1 = \frac{1}{16} = a_{11}$, $b_2 = \frac{1}{17} = a_{12}$, and in general $b_n = a_{n+10}$. Thus by discarding the first ten terms of $\{a_n\}$ the subsequence $\{b_n\}$ is formed (and $\{a_n\}$ and $\{b_n\}$ have the same limit, namely 0).

Now consider the following example. Let $\{a_n\} = \{2n - 7\}$ and $\{b_n\} = \{(-1)^{n+1}\}$. The first terms of $\{a_n\}$ are $-5, -3, -1, 1, 3$, and 5, and the terms of $\{b_n\}$ are alternately 1 and -1. Since $b_1 = 1 = a_4$ and $b_2 = -1 = a_3$, obviously order is not preserved. This means that $\{b_n\}$ is *not* a subsequence of $\{a_n\}$, even though every term of $\{b_n\}$ is a term of $\{a_n\}$.

THEOREM 3-2

Suppose $\{a_n\}$ is everywhere increasing, and suppose that $\{b_n\}$ increases without bound *and* is a subsequence of $\{a_n\}$. Then $\{a_n\}$ increases without bound.

This theorem can be proved in a fashion similar to the one used for proving Theorem 3-1.

Two Special Families of Sequences Which Increase Without Bound

THEOREM 3-3(a)

All sequences of the form $\{n^p\}$, where p is any real number > 0, increase without bound.

Examples of such sequences are

$$\{a_n\} = \{n\},\ \{b_n\} = \{n^2\},\ \{c_n\} = \{n^3\},\ \{d_n\} = \{n^{3/2}\},$$
$$\{e_n\} = \{n^{1/2}\} = \{\sqrt{n}\},\ \{f_n\} = \{\sqrt[3]{n}\},\ \text{and}\ \{g_n\} = \{n^{2/3}\}.$$

Whenever $p = 1$, we have the basic sequence $\{a_n\} = \{n\}$, which we have assumed to be increasing without bound.

Whenever $p > 1$, we have sequences such as $\{b_n\} = \{n^2\}$, $\{c_n\} = \{n^3\}$, and $\{d_n\} = \{n^{3/2}\}$, and each such sequence could be proved to increase without bound by comparison with $\{a_n\} = \{n\}$, using Theorem 3-1. For example, if $p = 2$, we have $n^2 \geq n \leftrightarrow n \geq 1$, which is true for all natural numbers. Hence $\{b_n\} = \{n^2\}$, by comparison with $\{a_n\} = \{n\}$, increases without bound.

Whenever $0 < p < 1$, we have sequences such as $\{e_n\} = \{\sqrt{n}\}$, $\{f_n\} = \{\sqrt[3]{n}\}$, and $\{g_n\} = \{n^{2/3}\}$. We have already shown that $\{e_n\} = \{\sqrt{n}\}$ contains $\{a_n\} = \{n\}$ as a subsequence; therefore, since $\{e_n\} = \{\sqrt{n}\}$ is everywhere increasing, we know that it increases without bound, by Theorem 3-2.

Similarly, $\{f_n\} = \{\sqrt[3]{n}\}$ contains $\{a_n\} = \{n\}$ as a subsequence, since $f_1 = a_1$, $f_8 = 2 = a_2$, $f_{27} = 3 = a_3$, etc. Since $\{f_n\} = \{\sqrt[3]{n}\}$ is everywhere increasing, it increases without bound, by Theorem 3-2. (See Figure 14.) By similar reasoning we could show that $\{h_n\} = \{n^{1/q}\} = \{\sqrt[q]{n}\}$ increases without bound for any natural number $q \geq 2$.

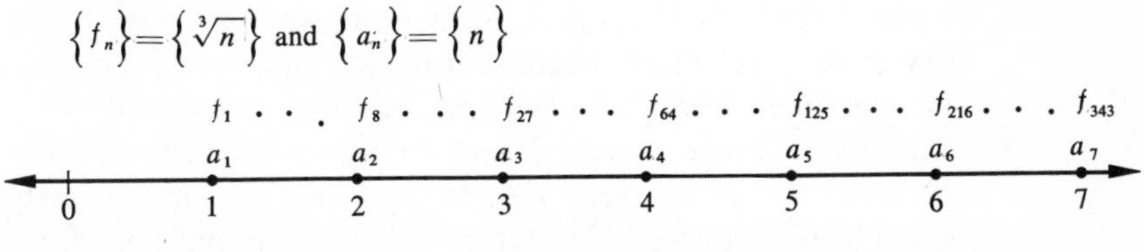

$\{f_n\} = \{\sqrt[3]{n}\}$ and $\{a_n\} = \{n\}$

$f_1 \ldots f_8 \ldots f_{27} \ldots f_{64} \ldots f_{125} \ldots f_{216} \ldots f_{343}$

$a_1 \quad a_2 \quad a_3 \quad a_4 \quad a_5 \quad a_6 \quad a_7$

0 1 2 3 4 5 6 7

Figure 14

Then $\{j_n\} = \{n^{r/q}\}$, where r and q are natural numbers ≥ 2 and $r < q$, can be written as $\{(n^{1/q})^r\}$, and Theorem 3-1 can be applied to show that $\{j_n\} = \{n^{r/q}\}$ increases without bound by comparison with $\{h_n\}$.

THEOREM 3-3(b)

All sequences of the form $\{p^n\}$, where p is any real number > 1, increase without bound.

Examples of such sequences are: $\{a_n\} = \{2^n\}$, $\{b_n\} = \{3^n\}$, $\{c_n\} = \{(\frac{5}{3})^n\}$, and

$$\{d_n\} = \left\{\left(1 + \frac{.04}{5}\right)^n\right\} = \{(1.008)^n\}.$$

We will not prove this theorem, but we will prove the special case in which $p = 2$, as follows.

Proof for the case in which $p = 2$

Compare the sequence $\{b_n\} = \{2^n\}$ with $\{a_n\} = \{n\}$, using logarithms to the base 2:

$$\begin{aligned} 2^n \geq n &\leftrightarrow \log_2 2^n \geq \log_2 n \\ &\leftrightarrow n \log_2 2 \geq \log_2 n \\ &\leftrightarrow \log_2 2 \geq \frac{\log_2 n}{n}. \end{aligned}$$

We have $\log_2 2 = 1$; and $1 \geq \frac{\log_2 n}{n}$ for all natural numbers, since the logarithm of any natural number is less than the number itself. For example, $\log_2 1 = \log_2 2^0 = 0$, $\log_2 2 = \log_2 2^1 = 1$, $\log_2 8 = \log_2 2^3 = 3$, etc. Thus by the Comparison Principle $\{2^n\}$ increases without bound.

Example Suppose \$1.00 is invested at 4% interest, compounded annually for an indefinite period of time. Let a_n be the amount of the investment (principal plus interest) at the end of n years. Then $a_n = (1 + .04)^n = (1.04)^n$. In approximately 18 years the accumulated interest would be equal to the initial investment, according to a Table of Compound Interest. That is, a_{18} would be approximately 2. If this point is not clear, we can simply realize that even at 4% simple interest, at the end of 25 years the accumulated interest would be equal to the initial investment. Since $\{a_n\}$ is everywhere increasing, and since the subsequence of $\{a_n\}$ obtained by taking every 18th term of $\{a_n\}$ is approximately equal to the sequence $\{2^n\}$, which increases without bound, we should suspect very strongly that $\{a_n\}$ increases without bound.

Applications of Theorems 3-1, 3-2, 3-3

Many additional sequences can be shown to increase without bound by either of the following methods:

(i) Comparing the given sequence with a sequence which is known to increase without bound, particularly a sequence of the form $\{p^n\}$ $(p > 1)$ or $\{n^p\}$ $(p > 0)$, thus using Theorem 3-1.

(ii) Showing that a given sequence which is everywhere increasing contains, as a subsequence, a sequence which is known to increase without bound, particularly a sequence of the form $\{p^n\}$ $(p > 1)$ or $\{n^p\}$ $(p > 0)$, thus using Theorem 3-2.

Examples **(1)** $\{b_n\} = \{\sqrt{n+5}\}$ increases without bound, by comparison with $\{a_n\} = \{\sqrt{n}\}$, since $\sqrt{n+5} \geq \sqrt{n} \leftrightarrow n + 5 \geq n$, which is true for all natural numbers.

(2) $\{b_n\} = \{\sqrt{n-1}\}$ increases without bound by comparison with $\{a_n\} = \{\sqrt[4]{n}\}$, since

$$\begin{aligned}\sqrt{n-1} \geq \sqrt[4]{n} &\leftrightarrow n - 1 \geq (\sqrt[4]{n})^2 = \sqrt{n}\\ &\leftrightarrow (n-1)^2 \geq (\sqrt{n})^2 = n\\ &\leftrightarrow n^2 + 1 \geq 3n\\ &\leftrightarrow n + \frac{1}{n} \geq 3, \text{ which is true for } n \geq 3.\end{aligned}$$

EXERCISES

1. Which sequences increase without bound?

$\{a_n\} = \{\sqrt{n^2 - n}\}$ $\qquad$ $\{b_n\} = \{2n^2 - \sin n\}$

$\{c_n\} = \left\{\left(2n - \frac{7}{n}\right)^3\right\}$ $\qquad$ $\{d_n\} = \left\{\sqrt[3]{\frac{n^2+1}{n}}\right\}$

2. (a) Which sequences in the Exercises of the preceding section (page 52) increase without bound? Are any of these sequences convergent?

(b) Which sequences in the Exercises of the preceding section appear to be convergent? Do any of these increase without bound?

3. Convergent Increasing Sequences

In the preceding section we discussed sequences which are everywhere increasing and unbounded above, hence increasing without bound and divergent. We shall now discuss sequences which are everywhere increasing and bounded above; such sequences are convergent, as we shall prove.

First we need to recognize a basic property of the system of real numbers.

LEAST UPPER BOUND AXIOM

Let S be any set of real numbers (not necessarily the terms of a sequence) which is bounded above. Of all the upper bounds of set S, one of these is less than any other upper bound of S and is called the *least upper bound* for S (abbreviated LUB).

For many number systems, such as the rational numbers and the decimal fractions (the finite decimal system), the least upper bound axiom does not hold true. Several such systems are discussed in the Appendix. It is highly recommended that you read this discussion in the Appendix very carefully before you complete this chapter *and again* when you have completed the last chapter of the book.

Since we are working within the real number system, we can assume the least upper bound axiom. On the basis of this axiom we shall prove the following theorem.

THEOREM 3-4

If $\{a_n\}$ is a sequence, with all terms real numbers, such that $\{a_n\}$ is bounded above and everywhere increasing, then

(1) $\{a_n\}$ converges, and

(2) the limit of $\{a_n\}$ is its least upper bound.

Proof

Let L be the least upper bound for $\{a_n\}$. Let $\langle L - E, L + E\rangle$ be any neighborhood of L. Since L is the least upper bound for $\{a_n\}$, any number less than L is not an upper bound. Hence $L - E$ is not an upper bound, so that there is at least one term of $\{a_n\}$ which is greater than $L - E$; call this term a_M. Thus $L - E < a_M$, and it is also true that $a_M \leq L$, since L is an upper bound. That is, $L - E < a_M \leq L$. Since $\{a_n\}$ is everywhere increasing, all terms a_n with $n > M$ are $> L - E$ and are also $\leq L$. Thus all terms a_n with $n \geq M$ are in $\langle L - E, L]$, and hence in $\langle L - E, L + E\rangle$. Thus by definition of convergence $\{a_n\} \to L$. (See Figure 15.)

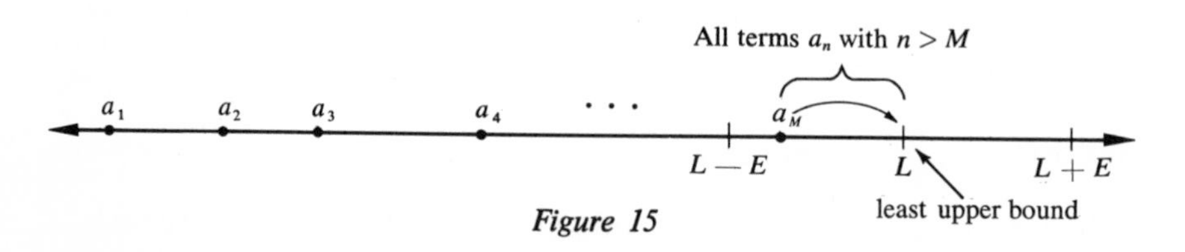

Figure 15

Example Show that $\{a_n\} = \left\{\dfrac{5n}{3n + 1}\right\}$ is everywhere increasing and bounded above, and hence convergent by Theorem 3-4:

$$\begin{aligned} a_n \leq a_{n+1} &\leftrightarrow \frac{5n}{3n+1} \leq \frac{5(n+1)}{3(n+1)+1} \\ &\leftrightarrow \frac{5n}{3n+1} \leq \frac{5n+5}{3n+4} \\ &\leftrightarrow 5n(3n+4) \leq (3n+1)(5n+5) \\ &\leftrightarrow 15n^2 + 20n \leq 15n^2 + 20n + 5 \\ &\leftrightarrow 0 \leq 5. \end{aligned}$$

Thus $a_n \leq a_{n+1}$ is true for all natural numbers, so $\{a_n\}$ is everywhere increasing. (See Figure 16.)

An investigation of several terms of the sequence may prompt us to select 2 as a possible upper bound. Let us test 2, as follows:

$$\frac{5n}{3n+1} \leq 2 \leftrightarrow 5n \leq 6n + 2$$

$$\leftrightarrow 0 \leq n + 2, \text{ which is true for all natural numbers } n.$$

Thus 2 is an upper bound (but not necessarily the least upper bound). Or, we could have tested 3 and found that it is an upper bound. Thus we have shown that $\{a_n\} = \left\{\frac{5n}{3n+1}\right\}$ is everywhere increasing and bounded above. Consequently, by Theorem 3-4, the sequence is convergent.

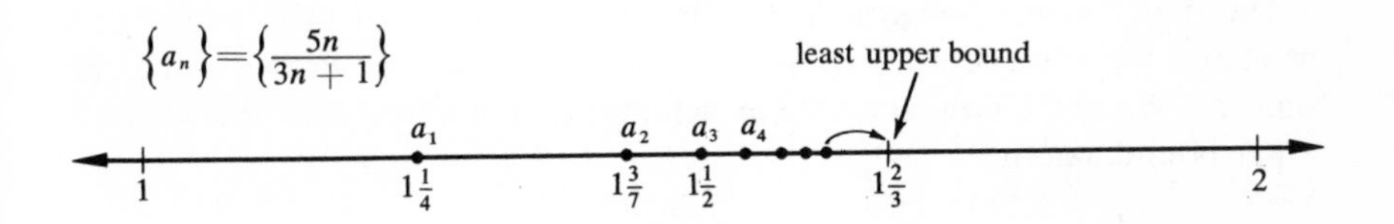

Figure 16

We still do not know, by using this theorem, what the limit is. To actually find the limit we would have to employ other means. We could divide both numerator and denominator of the general term by n, obtaining $\dfrac{5}{3 + \frac{1}{n}}$ and guess that the limit is $\frac{5}{3}$. It so happens that this guess is accurate, as we could demonstrate by using the definition of convergence studied in Chapter 2. Possibly we are not even interested in knowing what number the limit is; knowing that the limit does exist may be sufficient in certain situations.

Significance of Theorem 3-4

You are concerned in this book with understanding what it means to say "A sequence $\{a_n\}$ converges." The very basic meaning of convergence was indicated in Chapter 2 by the definition of convergence, which states as follows: A sequence $\{a_n\}$ is said to converge to a number A if, for every neighborhood of A, there can be found a natural number M such that all terms of $\{a_n\}$ with $n \geq M$ are in the neighborhood.

This definition itself can often be used without undue difficulty to prove that a given sequence converges. In many instances, however, the definition is very inconvenient to use for actually proving convergence because the necessary inequalities may be extremely difficult to solve and because the definition requires that we correctly guess the limit beforehand. In Chapter 2, you perhaps wondered at times how you would prove that a certain sequence converges if you were unsuccessful in guessing the limit.

Theorem 3-4 provides a means for proving that a sequence converges which does not require that we guess the suspected limit. To apply Theorem 3-4, we have to prove that the sequence is everywhere increasing, and we have to find an upper bound.

Throughout the remainder of this book, you will become acquainted with still other means for showing convergence. In fact, one of our principal objectives in studying sequences is to arrive at certain rapid techniques for showing that a sequence converges and for finding its limit.

4. Decreasing Sequences

The major ideas discussed in the first three sections of this chapter can be stated for sequences which are *everywhere decreasing*. Proofs would be similar. We shall now state these definitions, theorems, and the greatest lower bound axiom.

DEFINITIONS

A sequence $\{a_n\}$ is said to be *everywhere decreasing* if no term is less than the succeeding term — that is, for every natural number n, $a_n \geq a_{n+1}$.

A sequence $\{a_n\}$ is said to be *bounded below* if there is a number B such that no term of the sequence is less than B. That is, for all natural numbers n, $a_n \geq B$. Such a number is called a *lower bound* for the sequence.

If a sequence $\{a_n\}$ is not bounded below, then $\{a_n\}$ is said to be *unbounded below*.

If a sequence is everywhere decreasing and is unbounded below, then the sequence is said to *decrease without bound*.

GREATEST LOWER BOUND AXIOM

Let S be any set of real numbers (not necessarily the terms of a sequence) which is bounded below. Of all the lower bounds of set S, one of these is greater than any other lower bound of S and is called the *greatest lower bound* for S (abbreviated GLB).

THEOREM 3-5

Comparison Principle. Let $\{a_n\}$ be a sequence which decreases without bound and let $\{b_n\}$ be a sequence which is everywhere decreasing. If there is a particular natural number M such that, for all natural numbers $n \geq M$, $b_n \leq a_n$, then $\{b_n\}$ also decreases without bound.

THEOREM 3-6

Suppose $\{a_n\}$ is everywhere decreasing and suppose that $\{b_n\}$ decreases without bound *and* is a subsequence of $\{a_n\}$. Then $\{a_n\}$ decreases without bound.

THEOREM 3-7

If $\{a_n\}$ is a sequence, with all terms real numbers, such that $\{a_n\}$ is bounded below and everywhere decreasing, then $\{a_n\}$ converges, and the limit of $\{a_n\}$ is its greatest lower bound.

EXERCISES

1. Determine which of the sequences whose general term is given are convergent by Theorem 3-4 or by Theorem 3-7. If the sequence does converge, guess the limit. (You need not prove that your guess is correct.)

$$a_n = \frac{2n+3}{n+4} \qquad g_n = \frac{n}{3^n}$$

$$b_n = \frac{8n}{7n+20} \qquad j_n = 3 - \frac{1}{n^2}$$

$$c_n = \begin{cases} -3 \text{ for } n \text{ odd} \\ -3 - \dfrac{1}{(n+1)^2} \text{ for } n \text{ even} \end{cases} \qquad k_n = \frac{(n+1)(n+2)}{n(n+3)}$$

$$p_n = 2^{1/n}$$

$$d_n = (-1)^n\left(2 + \frac{1}{n}\right)$$

2. If $\{a_n\}$ increases without bound, what can be said about convergence of $\left\{\dfrac{1}{a_n}\right\}$?

5. More About Increasing and Decreasing Sequences

For the sequence $\{c_n\} = \left\{\frac{n^2}{2^n}\right\}$ the inequality $c_n \geq c_{n+1}$ can be solved as follows:

$$
\begin{aligned}
c_n \geq c_{n+1} &\leftrightarrow \frac{n^2}{2^n} \geq \frac{(n+1)^2}{2^{n+1}} \\
&\leftrightarrow \frac{2(n^2)}{2^{n+1}} \geq \frac{(n+1)^2}{2^{n+1}} \\
&\leftrightarrow 2(n^2) \geq (n+1)^2 \\
&\leftrightarrow 2n^2 \geq n^2 + 2n + 1 \\
&\leftrightarrow n \geq 2 + \frac{1}{n}, \text{ which is true for all } n \geq 3.
\end{aligned}
$$

Thus $c_3 \geq c_4$, $c_4 \geq c_5$, $c_5 \geq c_6$, and in general for n any natural number ≥ 3, $c_n \geq c_{n+1}$. This means that $\{c_n\}$ is *decreasing beginning with the third term.* Furthermore the inequality $c_n \leq c_{n+1}$ is equivalent to $n \leq 2 + \frac{1}{n}$ and is true for $n = 1$ and 2, so that $\{c_n\}$ is *increasing for the first two terms.*

Since $\{c_n\}$ is decreasing for $n \geq 3$ and since the sequence is bounded below by 0, then the sequence must converge. Try to guess its limit! (See Figure 17. Complete the graph yourself.)

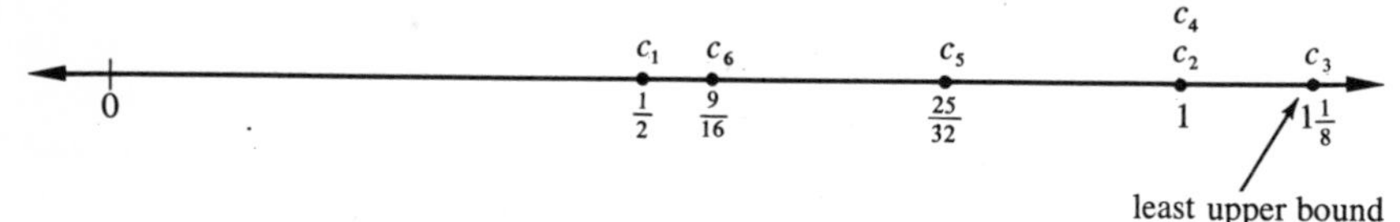

Figure 17. First Six Terms of $\{c_n\} = \left\{\frac{n^2}{2^n}\right\}$

Let us state the following pertinent definitions and theorems. Proofs of these two theorems are not included.

DEFINITIONS

Let $\{a_n\}$ be a sequence and M a natural number.

(i) If the inequality $a_n \leq a_{n+1}$ is true for all natural numbers $n \geq M$, then $\{a_n\}$ is said to be *increasing beginning with the Mth term.* If $M = 1$, $\{a_n\}$ is said to be *everywhere increasing.*

(ii) If $a_n \leq a_{n+1}$ is true for all natural numbers $n \leq M$, then $\{a_n\}$ is said to be *increasing for the first M terms.*

(iii) If $a_n \geq a_{n+1}$ is true for all natural numbers $n \geq M$, then $\{a_n\}$ is said to be *decreasing beginning with the Mth term.* If $M = 1$, $\{a_n\}$ is said to be *everywhere decreasing.*

(iv) If $a_n \geq a_{n+1}$ is true for all natural numbers $n \leq M$, then $\{a_n\}$ is said to be *decreasing for the first M terms.*

THEOREM 3-8

If $\{a_n\}$ is a sequence which is bounded above and which is increasing beginning with some term other than the first term, then $\{a_n\}$ converges. (Its limit might or might not be its least upper bound.)

THEOREM 3-9

If $\{a_n\}$ is a sequence which is bounded below and which is decreasing beginning with some term other than the first term, then $\{a_n\}$ converges. (Its limit might or might not be its greatest lower bound.)

The ideas which we have uncovered so far, relating to increasing sequences, decreasing sequences, upper bounds, lower bounds, etc., can often be very helpful in investigating the behavior of the terms of a sequence. Graphing a sequence on the real number line can be facilitated greatly by possession of such information, as we shall point out by several examples and exercises.

Example (1) Let $\{a_n\} = \left\{\dfrac{5n + 7}{2n}\right\}$. The first six terms are 6, $4\frac{1}{4}$, $3\frac{2}{3}$, $3\frac{3}{8}$, $3\frac{1}{5}$ and $3\frac{1}{12}$. To determine for which terms the sequence is decreasing or increasing we solve the inequality $a_n \geq a_{n+1}$ as follows:

$$a_n \geq a_{n+1} \leftrightarrow \frac{5n + 7}{2n} \geq \frac{5(n + 1) + 7}{2(n + 1)}$$

$$\leftrightarrow \frac{5n + 7}{2n} \geq \frac{5n + 12}{2n + 2}$$

$$\leftrightarrow (5n + 7)(2n + 2) \geq (2n)(5n + 12)$$

$$\leftrightarrow 10n^2 + 24n + 14 \geq 10n^2 + 24n$$

$$\leftrightarrow 14 \geq 0.$$

Thus the inequality $a_n \geq a_{n+1}$ is true for all natural numbers, and $\{a_n\}$ is everywhere decreasing.

Also, $\{a_n\}$ is bounded below, by 0, since all terms are positive. Therefore, since $\{a_n\}$ is everywhere decreasing and bounded below, it converges.

At this point in our investigation of this sequence, we could employ one or more of several tactics:

(1) We could search for a lower bound which is greater than 0. For example, we could test 2 as a lower bound by solving the inequality $2 \leq \dfrac{5n+7}{2n}$, obtaining $4n \leq 5n + 7$ and $0 \leq n + 7$, which is true for all natural numbers. This means that 2 is a lower bound.

We might suspect that 3 is a lower bound and test this number by solving the inequality $3 \leq \dfrac{5n+7}{2n}$. Doing so shows that only the first seven terms are ≥ 3.

Now we know that the limit (which is the greatest lower bound) is ≥ 2 and < 3, and we know that all terms except the first seven terms lie in the interval $[2, 3\rangle$. Thus we have the graph shown in Figure 18a.

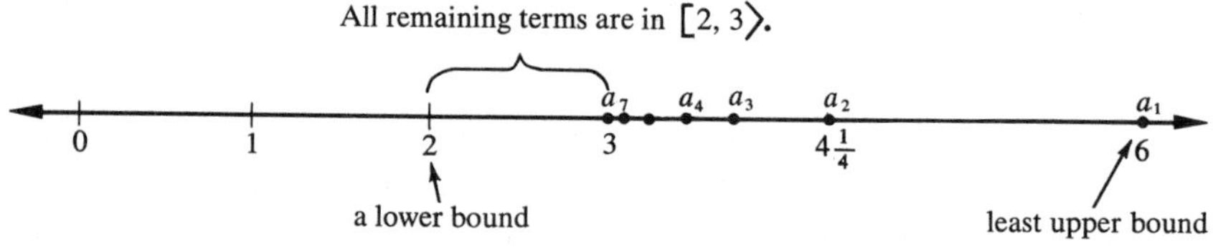

Figure 18a. First Graph of $\{a_n\} = \left\{\dfrac{5n+7}{2n}\right\}$

(2) We could write the general term $\dfrac{5n+7}{2n}$ as $\dfrac{5n}{2n} + \dfrac{7}{2n} = 2\dfrac{1}{2} + \dfrac{7}{2n}$ and observe that as n increases the values of $\dfrac{7}{2n}$ appear to approach 0 as a limit. This indicates that the limit is probably $2\frac{1}{2}$, and we could actually prove this by using the definition of convergence.

(3) We could compute a term of the sequence for n very large, say 10,000. Then $a_{10,000} = \dfrac{50{,}007}{20{,}000}$ which is very close to $\frac{5}{2} = 2\frac{1}{2}$. This also indicates that the limit is probably $2\frac{1}{2}$. (See Figure 18b.)

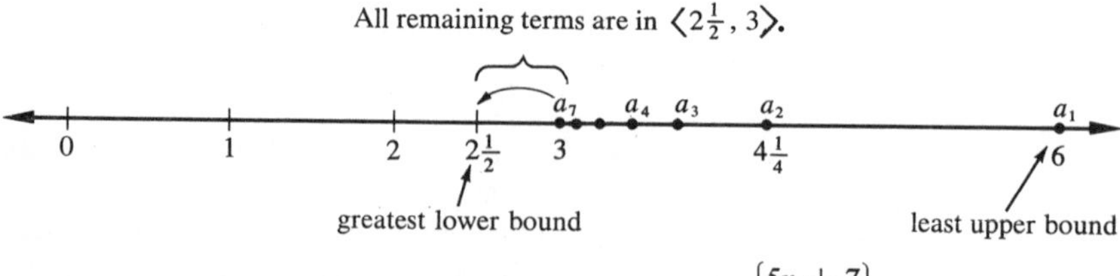

Figure 18b. Second Graph of $\{a_n\} = \left\{\dfrac{5n+7}{2n}\right\}$

Example (2) Let $\{b_n\} = \{16n - n^2\} = \{n(16 - n)\}$. The first 18 terms are 15, 28, 39, 48, 55, 60, 63, 64, 63, 60, 55, 48, 39, 28, 15, 0, -17, and -36. The inequality $b_n \geq b_{n+1}$ is solved as follows:

$$\begin{aligned} b_n \geq b_{n+1} &\leftrightarrow 16n - n^2 \geq 16(n+1) - (n+1)^2 \\ &\leftrightarrow 16n - n^2 \geq 16n + 16 - n^2 - 2n - 1 \\ &\leftrightarrow 0 \geq 15 - 2n \\ &\leftrightarrow n \geq 8. \end{aligned}$$

Thus the sequence is *decreasing beginning with the eighth term.* (See Figure 19a.) The sequence diverges by comparison with $\{-n\}$:

$$\begin{aligned} 16n - n^2 \leq -n &\leftrightarrow 17n \leq n^2 \\ &\leftrightarrow 17 \leq n. \end{aligned}$$

Thus beginning with the 17th term and continuing thereafter, $b_n \leq -n$, so that the sequence $\{b_n\}$ diverges. (Recall the discussion on Divergent Sequences in Chapter 2, Section 8.)

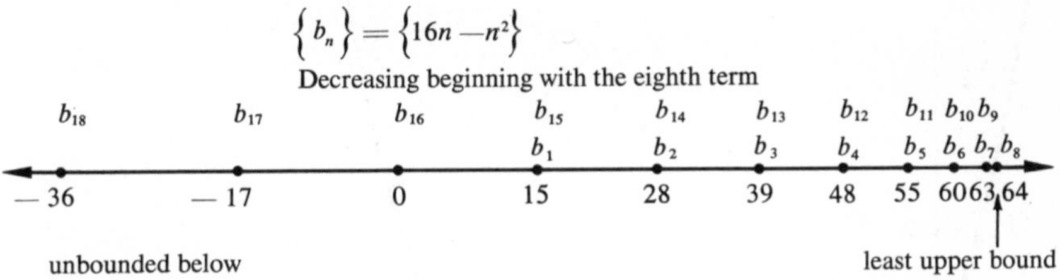

$\{b_n\} = \{16n - n^2\}$

Decreasing beginning with the eighth term

b_{18} b_{17} b_{16} b_{15} b_{14} b_{13} b_{12} b_{11} b_{10} b_9

b_1 b_2 b_3 b_4 b_5 b_6 b_7 b_8

-36 -17 0 15 28 39 48 55 60 63 64

unbounded below

least upper bound

Figure 19a

Example (3) Let $\{c_n\} = \left\{(-1)^n \left(\dfrac{n}{n+1}\right)\right\}$. The expression $(-1)^n$ prevents our solving the inequality $c_n \geq c_{n+1}$ by usual techniques. However, it is easily observed that the terms of this sequence are neither increasing nor decreasing. Rather, they are alternating from positive to negative. (See Figure 19b.)

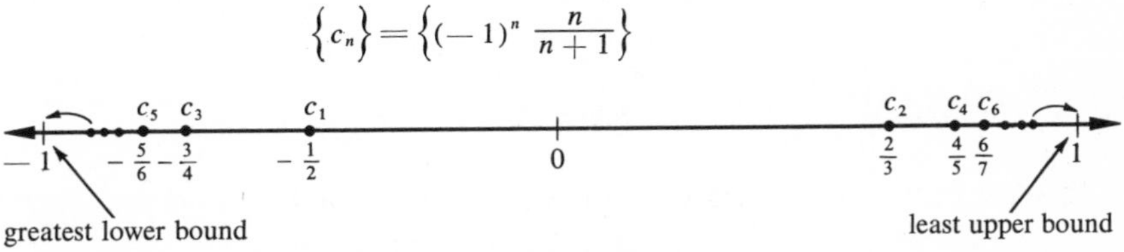

$\{c_n\} = \left\{(-1)^n \dfrac{n}{n+1}\right\}$

c_5 c_3 c_1 c_2 c_4 c_6

-1 $-\frac{5}{6}$ $-\frac{3}{4}$ $-\frac{1}{2}$ 0 $\frac{2}{3}$ $\frac{4}{5}$ $\frac{6}{7}$ 1

greatest lower bound

least upper bound

Figure 19b

REMARK: In the following exercises you will sometimes encounter certain inequalities of the form $P(n) \geq 0$, where $P(n)$ is some quadratic polynomial in the variable n. For example, $P(n)$ may be $2n^2 - 23n - 12$. Then we will need to factor $2n^2 - 23n - 12$, obtaining $(2n + 1)(n - 12) \geq 0$. Since n is replaceable only by natural numbers, $2n + 1$ is always positive and $n - 12$ is ≥ 0 whenever $n \geq 12$. Thus $2n^2 - 23n - 12 \geq 0$ is true whenever $n \geq 12$.

Another way to solve the inequality $2n^2 - 23n - 12 \geq 0$ is to graph the quadratic function $f(x) = 2x^2 - 23x - 12$ and determine the values of x for which the graph lies on or above the x-axis. Then the natural numbers in this set will be the solution set for the inequality $2n^2 - 23n - 12 \geq 0$. For $f(x) = 2x^2 - 23x - 12 = (2x + 1)(x - 12)$, $f(x) \geq 0$ for $x \leq -\frac{1}{2}$ or $x \geq 12$. The natural numbers in this set are the natural numbers ≥ 12. (See Figure 20.)

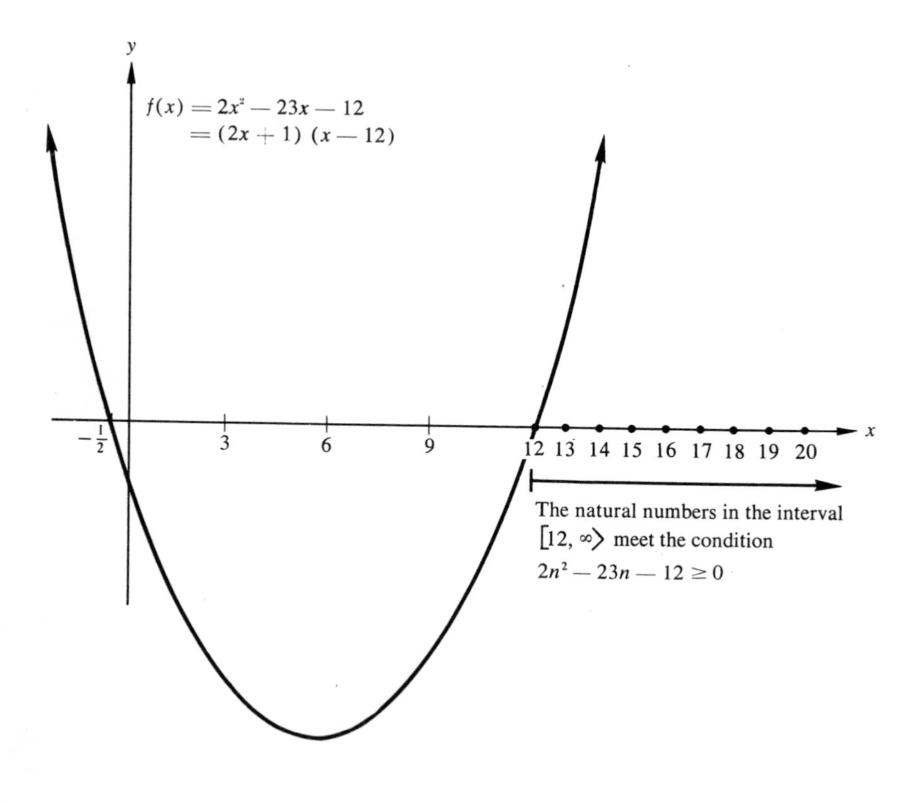

Figure 20

In some instances we need to use the quadratic formula

$$x = \frac{-b \pm \sqrt{b^2 - 4ac}}{2a}$$

to find the values of x for which $f(x) = 0$ — that is, the points at which the graph of $y = f(x)$ comes into contact with the x-axis — provided that such values exist.

EXERCISES

For each of the sequences whose general term is given (in 1–10), do the following:

(a) Determine for which terms the sequence is increasing and for which terms it is decreasing.

(b) Graph the sequence.

(c) Find an upper bound and a lower bound for the sequence, whenever they exist (not necessarily the *least* upper bound or the *greatest* lower bound).

(d) Decide whether or not the sequence converges. If it converges, guess the limit.

1. $a_n = n^2 - 10n$

2. $b_n = \dfrac{2^n}{3^{n-1}}$

3. $c_n = \dfrac{n^2 + 7}{n^2 + n}$

4. $d_n = \dfrac{n^2}{n + 10}$

5. $f_n = \dfrac{9n - 14}{n^2}$

6. $g_n = \dfrac{3n^2 + n}{n^2 + 10}$

7. $h_n = \dfrac{n^2 + 3n}{n^2 + 15}$

8. $j_n = \sin \dfrac{n}{2}\pi + \cos n\pi$

9. $k_n = \dfrac{1}{\sqrt{n+1} - \sqrt{n}}$

10. $p_n = \dfrac{n^2 - n}{2^n}$

11. Which of the sequences in Problems 1–10 converge to a number which is neither the LUB nor the GLB for the sequence?

12. The sequence $\{c_n\} = \left\{\dfrac{n^2 + 7}{n^2 + n}\right\}$ in Problem 3 decreases for $1 \leq n \leq 14$ and increases for $n \geq 14$. For which terms are each of the following sequences increasing and for which terms are they decreasing?

$$\{a_n\} = \left\{\frac{n^2 + 5}{n^2 + n}\right\} \qquad \{s_n\} = \left\{\frac{n^2 + n}{n^2 + 4}\right\}$$

$$\{b_n\} = \left\{\frac{n^2 + n}{n^2 + 5}\right\}$$

13. Does $\{t_n\} = \left\{\dfrac{1}{\sqrt{n+1} - \sqrt{n}}\right\}$ converge or diverge? Prove your answer.

$$\left[\text{Hint: Recall that } \frac{1}{a - b} \cdot \frac{a + b}{a + b} = \frac{a + b}{a^2 - b^2}.\right]$$

14. (a) If a sequence is everywhere decreasing, show that the sequence is bounded above.
(b) If a sequence is everywhere increasing, show that the sequence is bounded below.

15. Find, if possible, a sequence which is both everywhere increasing and everywhere decreasing.

16. Assuming that $\{\log n\}$ is everywhere increasing and unbounded above and that $\log 1 = 0$, show that $\{a_n\}$ is everywhere increasing, is bounded above by 1, and converges to 1, where $a_n = \dfrac{\log n}{\log 2n}$.

17. Either prove each of the following theorems or disprove it by finding a sequence for which it does not hold true:
(a) If a sequence is convergent, then it is bounded below and above.
(b) If a sequence is bounded below and above, then it is convergent.

6. Convergence of Geometric Sequences

Geometric sequences probably are not new to you. They were discussed briefly in the last section of Chapter 1. No doubt you studied them in a previous course. Whether or not they are new to you is somewhat unimportant at this time, since we wish to consider them within the context of limits of sequences. Our present intention is to investigate the convergence of various categories of geometric sequences. First, however, we shall review briefly the meaning of a geometric sequence.

A geometric sequence is a sequence whose general term can be placed in the form a_1r^{n-1}, where a_1 is the first term and r is the ratio of any term (after the first term) to its predecessor. The numbers a_1 and r may be positive or negative.

Examples of Geometric Sequences

General Term	*First Five Terms*
$3(\frac{1}{2})^{n-1}$ or $\dfrac{3}{2^{n-1}}$	$3, 1\frac{1}{2}, \frac{3}{4}, \frac{3}{8}, \frac{3}{16}$
$(-3)(\frac{1}{2})^{n-1}$ or $\dfrac{-3}{2^{n-1}}$	$-3, -1\frac{1}{2}, -\frac{3}{4}, -\frac{3}{8}, -\frac{3}{16}$
$(\frac{2}{3})(\frac{2}{3})^{n-1}$ or $(\frac{2}{3})^n$	$\frac{2}{3}, \frac{4}{9}, \frac{8}{27}, \frac{16}{81}, \frac{32}{243}$
$(-\frac{2}{3})(-\frac{2}{3})^{n-1}$ or $(-\frac{2}{3})^n$	$-\frac{2}{3}, \frac{4}{9}, -\frac{8}{27}, \frac{16}{8}, -\frac{32}{243}$
$2(2^{n-1})$ or 2^n	$2, 4, 8, 16, 32$
$(-1)(-1)^{n-1}$ or $(-1)^n$	$-1, 1, -1, 1, -1$

EXERCISES

1. Assume $a_1 = 1$. Then the geometric sequence $\{a_n\}$ has the form $\{r^{n-1}\}$. For which of the given values of r is $\{r^{n-1}\}$ (a) everywhere increasing? (b) everywhere decreasing? (c) convergent?

$r = -3$ $\quad r = -\frac{1}{2}$ $\quad r = 1$

$r = -1$ $\quad r = \frac{1}{2}$ $\quad r = 3$

2. Assume $a_1 = 1$. Determine for which of the following sets of values of r the geometric sequence $\{r^{n-1}\}$ converges. (There will be more than one set.) Then, determine which of those sets completely describes the values of r for which $\{r^{n-1}\}$ converges.

(a) $r < -1$

(b) $r \leq -1$

(c) $-1 < r < 0$

(d) $-1 \leq r < 0$

(e) $0 < r < 1$

(f) $0 < r \leq 1$

(g) $r > 1$

(h) $r \geq 1$

(i) $-1 < r < 0$ or $0 < r \leq 1$

(j) $0 < |r| < 1$ (that is, $-1 < r < 0$ or $0 < r < 1$)

(k) $0 < |r| \leq 1$ (that is, $-1 \leq r < 0$ or $0 < r \leq 1$)

3. Which of the sets (a)–(k) of Problem 2 completely describes the values of r for which $\{r^{n-1}\}$ is:

(a) everywhere increasing?

(b) everywhere decreasing?

4. Prove that if a_1 is some positive number other than 1, the values of r for which $\{a_1 r^{n-1}\}$ converges are exactly the same as those for which $\{r^{n-1}\}$ converges.

If a_1 is negative, do $\{a_1 r^{n-1}\}$ and $\{r^{n-1}\}$ converge for exactly the same values of r?

4 • *Applications: Rate of Change*

1. Preliminary Ideas

In this chapter we will discuss applications of limits of sequences to several physical problems involving the concept of rate of change. For instance, we will examine the meaning of velocity of a moving object as one particular kind of rate of change. We will also study the notion of slope of a curve at a given point and the related idea of a tangent line to a curve.

The purpose of this first section is to present certain preliminary ideas, most of which should be familiar to you from previous courses.

General Notion of a Function

DEFINITION

A *function* is a set of ordered pairs of real numbers in which no two pairs have the same first elements and different second elements. The set of all first elements is called the *domain* and the set of all second elements is called the *range*.

A function represents a relationship between the elements of two sets, called the domain and the range. By definition, every element in the domain corresponds to exactly one element in the range.

If the name of a function is f, then for any given element x in the domain the corresponding element in the range may be designated by $f(x)$. For example, if 2 is an element in the domain, then $f(2)$ — read "f of 2" or "f at 2" — is the corresponding element in the range, and $(2, f(2))$ is one of the ordered pairs comprising the function f.

If f is the function described by $f(x) = x^2 + 1$ and the domain is the set of all real numbers, then the range is the set $[1, \infty\rangle$, and $f(-2) = 5$, $f(2) = 5$, and $f(0) = 1$, for example.*

In our examples and exercises, if the domain of a given function is not specified, this will mean that the domain is the set of all real numbers for which the description of the function is meaningful. For example, let f and g be two functions described by $f(x) = x^2 + 2$ and $g(x) = \sqrt{x + 2}$, respectively. Then the domain of f is the set of all real numbers, and the domain of g is the set $[-2, \infty\rangle$ (that is, the set of all real numbers ≥ -2).

The graph of a function f is the set of all points in the coordinate plane represented by the ordered pairs $(x, f(x))$ of the function.

The following definitions will be useful in this chapter.

DEFINITIONS

Let I be an interval on the real number line; I may be open, closed, or neither. Let f be a function with I in its domain, and let x_1 and x_2 be any two numbers which are in the interval I.

If whenever $x_1 < x_2$ we have

(1) $f(x_1) \leq f(x_2)$,
then f is said to be *increasing in the interval I.*

(2) $f(x_1) < f(x_2)$,
then f is said to be *strictly increasing in the interval I.*

If whenever $x_1 < x_2$ we have

(1) $f(x_1) \geq f(x_2)$,
then f is said to be *decreasing in the interval I.*

(2) $f(x_1) > f(x_2)$,
then f is said to be *strictly decreasing in the interval I.*

REMARK: According to the preceding definitions, a constant function, say $f(x) = 2$, is increasing in any given interval and is also decreasing in any given interval. However, this function is not strictly increasing nor strictly decreasing in any interval. Later in this chapter, in our discussions of slopes of tangent lines, we will be concerned mainly with strictly increasing and strictly decreasing functions.

* It is important to keep in mind that the symbols ∞ and $-\infty$ do *not* represent numbers. For example, the expression $[1, \infty\rangle$ in the present example is simply a short hand way of denoting the set of all real numbers ≥ 1. You may wish to refer back to the discussion in Section 1 of Chapter 2 on the use of the symbols ∞ and $-\infty$.

If $f(x) = x^2$, this function is strictly increasing in any interval whose endpoints are both positive and is strictly decreasing in any interval whose endpoints are both negative.

If $f(x) = x^3$, this function is strictly increasing in any interval of real numbers. (See Figure 21.)

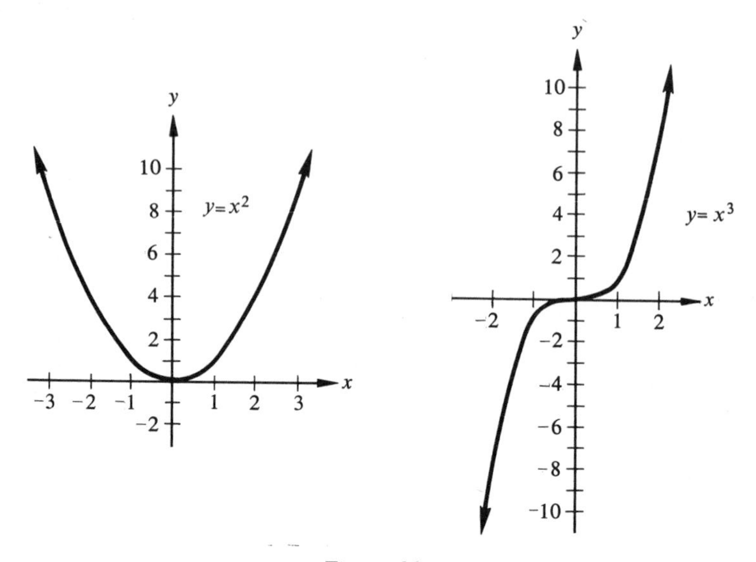

Figure 21

Slope of a Linear Function

A linear function is a function whose graph in the coordinate plane is a straight line. Suppose L is any line in the coordinate plane except a vertical or a horizontal line. Then, when examined from left to right, L will either slant upward, as represented in Figure 22a, or slant downward, as represented in Figure 22b.

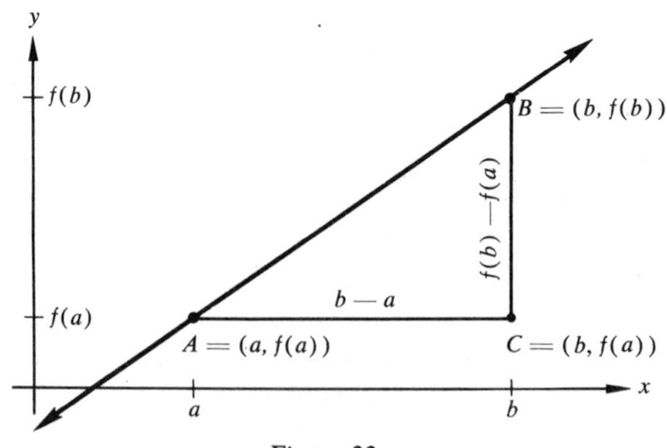

Figure 22a

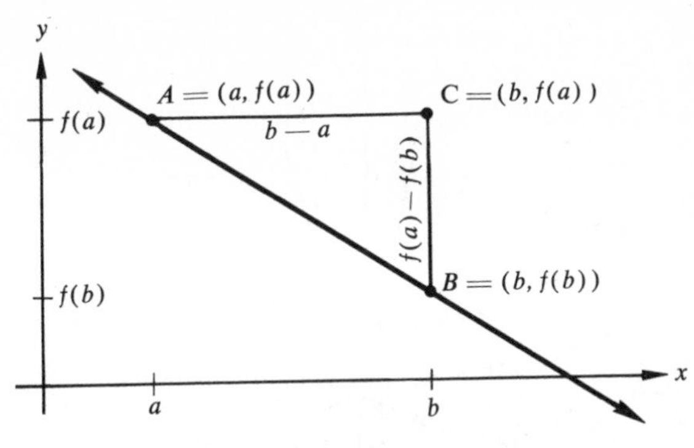

Figure 22b

Suppose that f is the name of the function whose graph is L. Let A and B be *any* two points on line L with coordinates $(a,f(a))$ and $(b,f(b))$, assuming that $a < b$. In each of the two cases (Figure 22a and Figure 22b) draw a line parallel to the x-axis through A and a line parallel to the y-axis through B. In both cases the two lines will intersect at right angles at a point $C = (b,f(a))$.

The following information pertaining to a linear function f will be useful for our purposes. We are not concerned here with proofs of these properties of a linear function. If desired, they are readily available from numerous other sources.

A linear function f can be described by using the form $f(x) = mx + d$, where m is the slope of the line and d is the point on the y-axis at which the y-axis and the graph of f intersect. The number m can be determined by the formula $m = \dfrac{f(b) - f(a)}{b - a}$, where $(a,f(a))$ and $(b,f(b))$ are any two points on the graph of f.

To illustrate the meaning of the concept of slope, we first consider Figure 22a. Since $b > a$ and $f(b) > f(a)$, then $b - a$ and $f(b) - f(a)$ are both positive, and therefore the ratio $m = \dfrac{f(b) - f(a)}{b - a}$ is positive. Geometrically, this ratio represents a comparison of the lengths of the line segments BC and AC.

However, in Figure 22b the ratio $m = \dfrac{f(b) - f(a)}{b - a}$ is negative because $b > a$ and $f(b) < f(a)$, so that $b - a > 0$ and $f(b) - f(a) < 0$.

In Figures 23a and 23b on the next page graphs of various linear functions are shown.

We now ask ourselves whether or not it is possible to have a linear function whose graph is parallel to one of the axes. Suppose L is parallel to the x-axis.

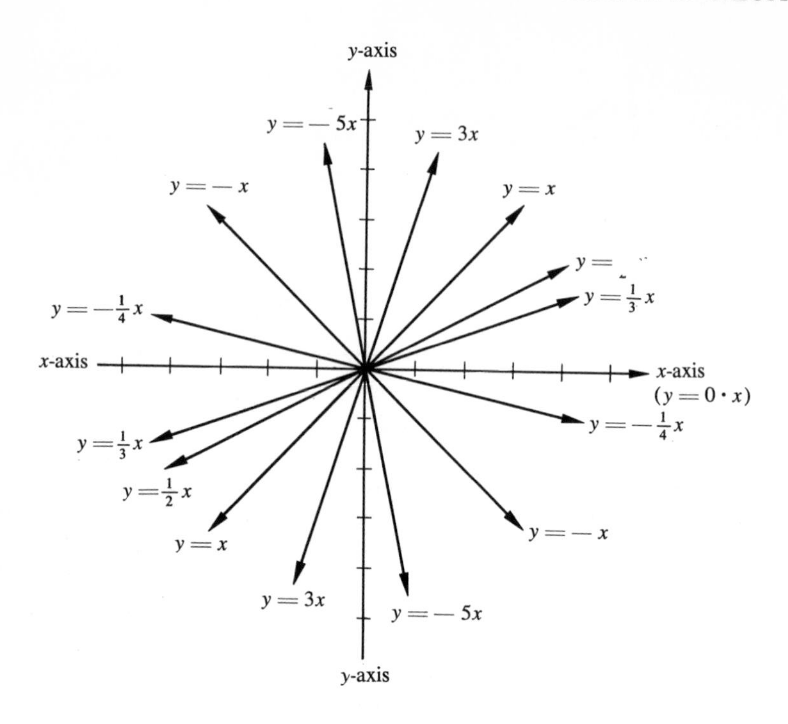

Figure 23a

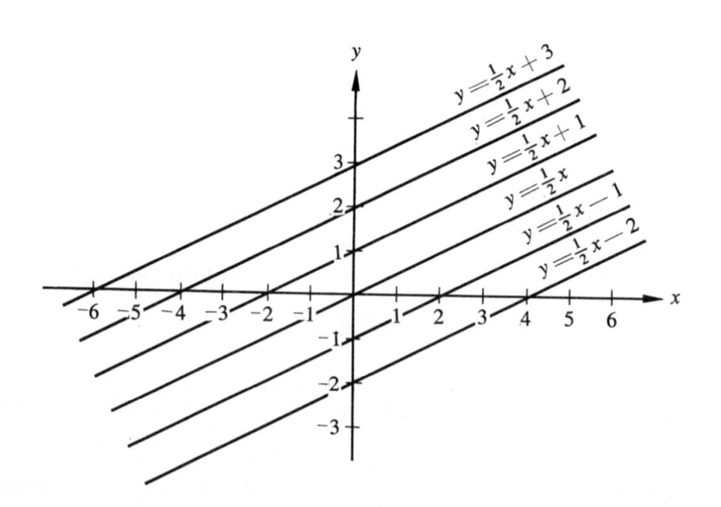

Figure 23b

If $A = (a, f(a))$ and $B = (b, f(b))$ are any two points on L, we have $f(a) = f(b)$, so that $\dfrac{f(b) - f(a)}{b - a} = \dfrac{0}{b - a} = 0$. That is, $m = 0$, and $f(x) = mx + d$ becomes $f(x) = d$. For example, if $d = 3$, we obtain the constant function $f(x) = 3$, whose graph is shown in Figure 24.

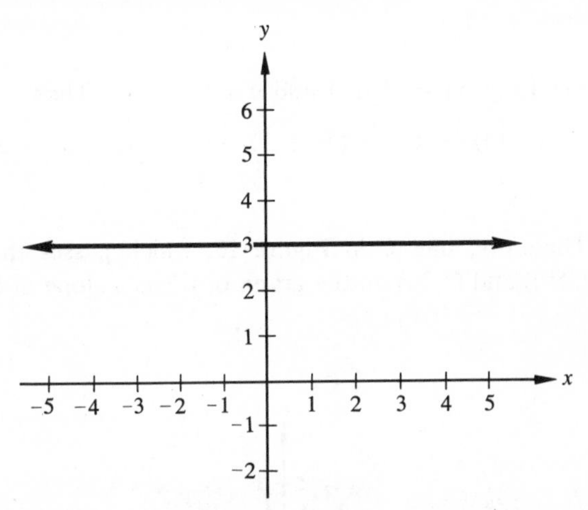

Figure 24. Graph of the Constant Function $f(x) = 3$

On the other hand, if L is vertical, then any two points on L would have the same x-coordinate and different y-coordinates. This means that we would not have the graph of a function, since, in order to have a function, any given element in the domain of a function must correspond to exactly one element in the range. Thus L is not the graph of a function whenever L is parallel to the y-axis.

Example **(1)** Find the function whose graph is the line containing the two points $(2, 5)$ and $(7, -2)$.

Solution: The slope $m = \dfrac{(-2) - (5)}{(7) - (2)} = -\frac{7}{5}$, so that $f(x) = -\frac{7}{5}x + d$. For $x = 2$, we have $f(2) = (-\frac{7}{5})(2) + d \leftrightarrow 5 = -\frac{14}{5} + d \leftrightarrow d = \frac{39}{5}$. (Or, we could let $x = 7$, obtaining $f(7) = (-\frac{7}{5})(7) + d \leftrightarrow -2 = -\frac{49}{5} + d \leftrightarrow d = \frac{39}{5}$.) Thus the desired function is described by $f(x) = -\frac{7}{5}x + \frac{39}{5}$.

Example **(2)** Find the function whose graph is the line which has slope of 3 and which contains the point $(2, 5)$.

Solution: Since $m = 3$, $g(x) = 3x + d$. For $x = 2$, we have $g(2) = (3)(2) + d \leftrightarrow 5 = 6 + d \leftrightarrow d = -1$. Thus the desired function is described by $g(x) = 3x - 1$.

Difference Quotients

Now suppose f is some function and $A = (a, f(a))$ and $B = (b, f(b))$ are any two points on the graph of f, with $a < b$. The slope of the line passing through A and B is the number $\frac{f(b) - f(a)}{b - a}$. The expression $\frac{f(b) - f(a)}{b - a}$ is called a *difference quotient.*

Example **(1)** Let $f(x) = x^2 + 1$ and $a = 3, b = 5$. Then

$$\frac{f(5) - f(3)}{5 - 3} = \frac{(5^2 + 1) - (3^2 + 1)}{5 - 3} = \frac{5^2 - 3^2}{2} = 8.$$

Therefore, line L in Figure 25, which passes through points (3,10) and (5,26) on the graph of f, has a slope of 8.

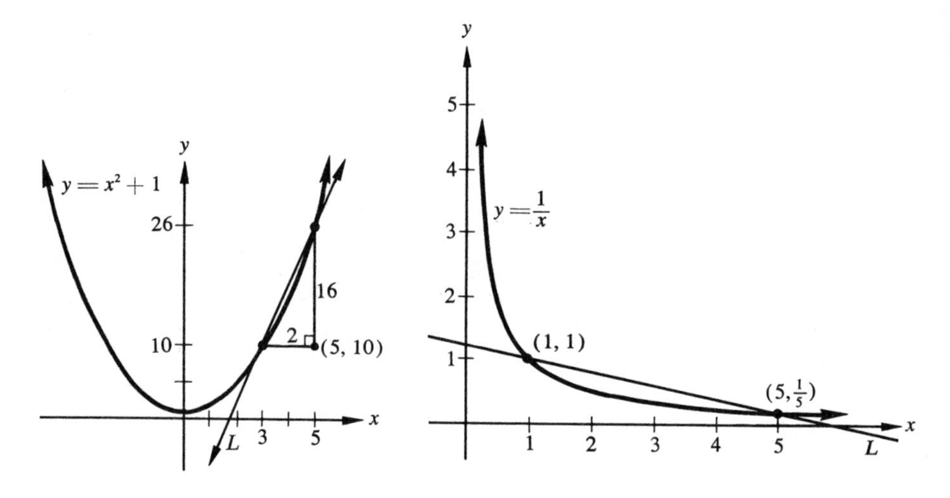

Figure 25 *Figure 26*

Example **(2)** Let $g(x) = \frac{1}{x}$, $a = 1$ and $b = 5$.

Then $\frac{g(5) - g(1)}{5 - 1} = \frac{\frac{1}{5} - 1}{4} = -\frac{1}{5}$. Therefore, the slope of line L in Figure 26 is $-\frac{1}{5}$. Since this slope is negative, the line L slants downward when examined from left to right. Furthermore, L is almost horizontal, since its slope is a number which is fairly close to 0.

Example (3) Let $h(x) = \sin x$ and $a = 1, b = 2$ (in radians).

Then $\dfrac{h(2) - h(1)}{2 - 1} = \dfrac{\sin 2 - \sin 1}{1}$, which is approximately $\dfrac{.9086 - .8415}{1} = .0671$. If $a = \dfrac{\pi}{2}$ and $b = \pi$, then $\dfrac{h(\pi) - h\left(\dfrac{\pi}{2}\right)}{\pi - \dfrac{\pi}{2}}$

$= \dfrac{\sin \pi - \sin \dfrac{\pi}{2}}{\dfrac{\pi}{2}} = \dfrac{0 - 1}{\dfrac{\pi}{2}} = -\dfrac{2}{\pi}$, which is approximately $-.63$, using $\pi = 3.14$.

Therefore, as we note in Figure 27, line K is almost horizontal, since its slope, approximately .0671, is very close to 0. Also, note that line L in this figure, with slope a negative number (approximately $-.63$), slants downward when examined from left to right.

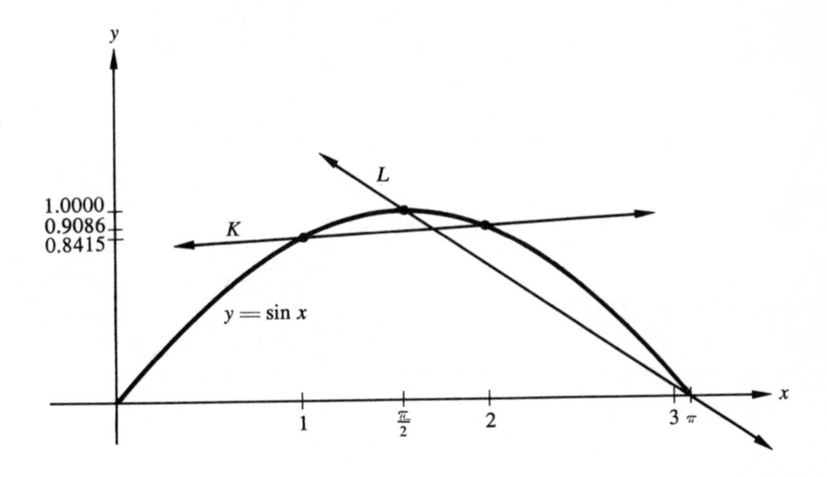

Figure 27

REMARK: Sometimes, instead of specifying one particular value for b when forming difference quotients, we specify a sequence whose terms are values for b. In this case, we select one particular value for a, and then we form a sequence of difference quotients. For example, let $a = 3$ and $b_n = 3 + \dfrac{1}{n}$, and let f be a function defined by $f(x) = x^2 + 1$.

Then the corresponding difference quotient is

$$\frac{f\left(3+\frac{1}{n}\right)-f(3)}{\left(3+\frac{1}{n}\right)-3} = \frac{\left[\left(3+\frac{1}{n}\right)^2+1\right]-[3^2+1]}{\frac{1}{n}}$$

$$= \frac{\left[9+\frac{6}{n}+\frac{1}{n^2}+1\right]-[9+1]}{\frac{1}{n}} = \frac{\frac{6}{n}+\frac{1}{n^2}}{\frac{1}{n}}$$

$$= 6+\frac{1}{n}.$$

Example (4) Let $g(x) = \frac{1}{x}$, $a = 2$, and $b_n = 2 + \frac{1}{n}$. Then the difference quotient is

$$\frac{g\left(2+\frac{1}{n}\right)-g(2)}{\frac{1}{n}} = \frac{\frac{1}{2+\frac{1}{n}}-\frac{1}{2}}{\frac{1}{n}} = \frac{\frac{2-\left(2+\frac{1}{n}\right)}{2\left(2+\frac{1}{n}\right)}}{\frac{1}{n}}$$

$$= \frac{\frac{-\frac{1}{n}}{2\left(2+\frac{1}{n}\right)}}{\frac{1}{n}} = n\left[\frac{-\frac{1}{n}}{2\left(2+\frac{1}{n}\right)}\right] = \frac{-1}{2\left(2+\frac{1}{n}\right)} = \frac{-1}{4+\frac{2}{n}}.$$

EXERCISES

1. Find the function whose graph is the line passing through each of the following pairs of points in the coordinate plane:

(a) (2, 1) and (5, 7)

(b) (5, 1) and (−3, 3)

(c) (7, 4) and (4, 7)

(d) (0, 0) and (5, 6)

(e) (0, 3) and (4, −5)

2. Find the function whose graph is the line which passes through the given point and which has the given slope:

(a) $(2, 6)$, $m = 4$
(b) $(2, 6)$, $m = -1$
(c) $(-2, -3)$, $m = 5$
(d) $(7, 4)$, $m = 3$

3. Simplify each difference quotient.

Function	*Difference Quotients*		
$f(x) = x^3 + 4$	$\dfrac{f(5) - f(3)}{5 - 3}$		
$g(x) = \dfrac{1}{x^2 + 1}$	$\dfrac{g(4) - g(3)}{4 - 3}$	and	$\dfrac{g(5) - g(3)}{5 - 3}$
$h(x) = x^2 + x$	$\dfrac{h(6) - h(4)}{6 - 4}$	and	$\dfrac{h\left(4 + \frac{1}{n}\right) - h(4)}{\frac{1}{n}}$
$k(x) = x^2 - 3x + 1$	$\dfrac{k(1) - k(-1)}{1 - (-1)}$	and	$\dfrac{k\left(2 + \frac{1}{n}\right) - k(2)}{\frac{1}{n}}$
$p(x) = \dfrac{x - 2}{x + 2}$	$\dfrac{p(6) - p(3)}{6 - 3}$	and	$\dfrac{p\left(6 + \frac{1}{n}\right) - p(6)}{\frac{1}{n}}$

2. Average Rate of Change

As an example, let us suppose an object (a small rock, perhaps) is dropped from a cliff which is 400 feet high, and suppose the distance (in feet) travelled by the object (in t seconds) is given by the function $f(t) = 16t^2$. Then exactly 5 seconds is required for the object to strike the ground directly below the cliff. Let us investigate the changes taking place in $f(t)$ as the value of t is changed.

As the value of t is increased from	the value of $f(t)$ is increased from	which is a change in distance of
0 to 1 sec.	0 to 16 ft.	16 ft.
1 to 2 secs.	16 to 64 ft.	48 ft.
2 to 3 secs.	64 to 144 ft.	80 ft.
3 to 4 secs.	144 to 256 ft.	112 ft.
4 to 5 secs.	256 to 400 ft.	144 ft.

Thus as we choose various unit intervals of time — intervals of length one second — the accompanying changes in distance traveled by the falling object will differ.

Now suppose we choose intervals of time which are not unit intervals, such as [0, 2], [0, 3], [2, 5], $[2\frac{1}{2}, 4]$, $[3\frac{2}{3}, 4]$. For such an interval we may wish to determine not simply the accompanying change in distance of the falling object, but rather the *average* rate of change per unit interval. For instance, in the time interval [0, 3], measured in seconds, the object falls 144 feet, an *average* of $\frac{144}{3}$ or 48 feet *per second.* In general, if $[a, b]$ is any interval which is a subset of the domain, [0, 5], then the average rate of change for that interval is given by the difference quotient $\frac{f(b) - f(a)}{b - a}\cdot$ Accordingly, for the time interval [1, 4] the average rate of change is

$$\frac{f(4) - f(1)}{4 - 1} = \frac{16(4)^2 - 16(1)^2}{3} = \frac{16}{3}[4^2 - 1^2] = \frac{16}{3}[15]$$

$$= 80 \text{ feet per second.}$$

Also, for the time interval $[3\frac{2}{3}, 4]$ the average rate of change is

$$\frac{f(4) - f(3\frac{2}{3})}{4 - 3\frac{2}{3}} = \frac{16(4)^2 - 16(3\frac{2}{3})^2}{\frac{1}{3}} = 3 \cdot 16[4^2 - (3\tfrac{2}{3})^2]$$

$$= 122\tfrac{2}{3} \text{ feet per second.}$$

Let us examine a graph of the function $f(t) = 16t^2$, with domain [0, 5], measured in seconds, and range [0, 400], measured in feet. (See Figure 28.) If $A = (a, f(a))$ and $B = (b, f(b))$ are any two points on the graph of f (with $a < b$), the average rate of change of f in the time interval $[a, b]$ is given by the difference quotient $\frac{f(b) - f(a)}{b - a}\cdot$ Moreover, the value of this difference quotient is the slope of the line passing through points A and B. In particular let

$A = (0, 0)$ $D = (3, 144)$
$B = (1, 16)$ $E = (4, 256)$
$C = (2, 64)$ $F = (5, 400)$

(See Figure 29.)

From the table on the opposite page we see that the average rate of change differs for various intervals within the domain of f. In the next section, we will return to this problem of the falling object and will determine what it means to speak of the rate of change of the object's position at a particular instant.

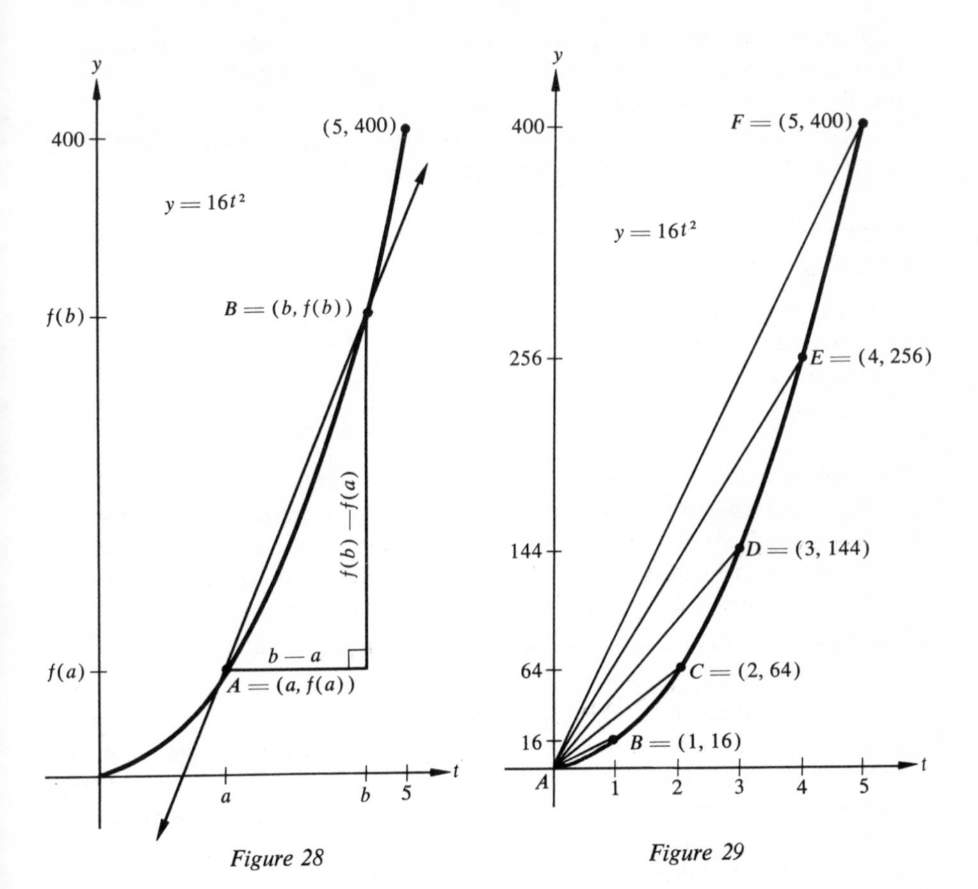

Figure 28 *Figure 29*

For the interval	the average rate of change is
[0, 1]	16 ft./sec.
[0, 2]	32 ft./sec.
[0, 3]	48 ft./sec.
[0, 4]	64 ft./sec.
[0, 5]	80 ft./sec.
[1, 2]	48 ft./sec.
[1, 3]	64 ft./sec.
[1, 4]	80 ft./sec.
[1, 5]	96 ft./sec.
[2, 5]	112 ft./sec.
[3, 5]	128 ft./sec.
[4, 5]	144 ft./sec.

It is now essential to state the following definitions and to consider examples (in addition to the preceding example) illustrating these definitions. The first definition is that of *average rate of change* of a function in a given interval $[a, b]$. Also defined are the particular notions of *average velocity* and *average speed* of an object moving in a straight line, during a given time interval.

DEFINITIONS

Let f be a function and let $[a,b]$ be an interval which is a subset of the domain of f. Then the *average rate of change* of f in $[a,b]$ is the number $\frac{f(b) - f(a)}{b - a}$.

Given a function f such that $f(t)$ describes the position at time t of an object moving in a straight line, for t in a given time interval $[a, b]$, then the *average velocity* of the object in $[a, b]$ is defined to be the average rate of change of the function f in $[a, b]$. Moreover, the *average speed* of the object in $[a, b]$ is the absolute value of its average velocity in that interval.

As the definitions point out, average velocity is a particular kind of average rate of change; the term *average velocity* can be used whenever the numbers in the domain of the function represent values of time and the numbers in the range denote the position, at each instant of time in the domain, of an object moving in a straight line. When the numbers in the domain do *not* refer to time, the more general term, *average rate of change*, should be used.

Suppose f is a function which describes the position at any instant of time of an object moving in a straight line. If f is strictly increasing in some time interval $[a, b]$, then $f(b) > f(a)$, and since $b > a$, then $\frac{f(b) - f(a)}{b - a}$ is positive. On the other hand, if the function is strictly decreasing in $[a, b]$, then $f(b) < f(a)$ and $\frac{f(b) - f(a)}{b - a}$ is negative. In either event, the average speed of the moving object in the interval $[a, b]$ is the absolute value of the average velocity, and consequently is positive.

Example **(1)** An object is thrown into the air in such a way that its distance (in feet) from the ground at any time t (in seconds) is given by the function $f(t) = t(6 - t)$. Find the average speed of the object in each of the following time intervals:

$$[0, 2], [2, 3], [3, 4], \text{ and } [4, 6].$$

Solution:

t	$f(t) = t(6 - t)$
0	0
2	8
3	9
4	8
6	0

$$\frac{f(2) - f(0)}{2 - 0} = 4 \text{ ft./sec.}$$

$$\frac{f(3) - f(2)}{3 - 2} = 1 \text{ ft./sec.}$$

$$\frac{f(4) - f(3)}{4 - 3} = -1 \text{ ft./sec.}$$

$$\frac{f(6) - f(4)}{6 - 4} = -4 \text{ ft./sec.}$$

The average speed of the object in [0, 2] and in [4, 6] is 4 ft./sec., and the average speed in [2, 3] and in [3, 4] is 1 ft./sec. Moreover, the average speed of the object in the interval [0, 3] is 3 ft./sec., and this is the same as the average speed in both the intervals [3, 6] and [0, 6].

Example **(2)** Suppose a 10-foot ladder is propped against a house with the lower end of the ladder x feet from the house. Then the distance in feet from the upper end of the ladder to the ground is $\sqrt{100 - x^2} = f(x)$. (See Figure 30.) Assume the lower end of the ladder is being pulled away from the house. Find the average rate of change of $f(x)$ as the value of x is increased from

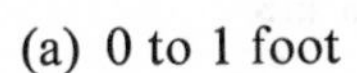

(a) 0 to 1 foot
(b) 1 to 2 feet
(c) 2 to 3 feet
(d) 6 to 8 feet
(e) 8 to 10 feet

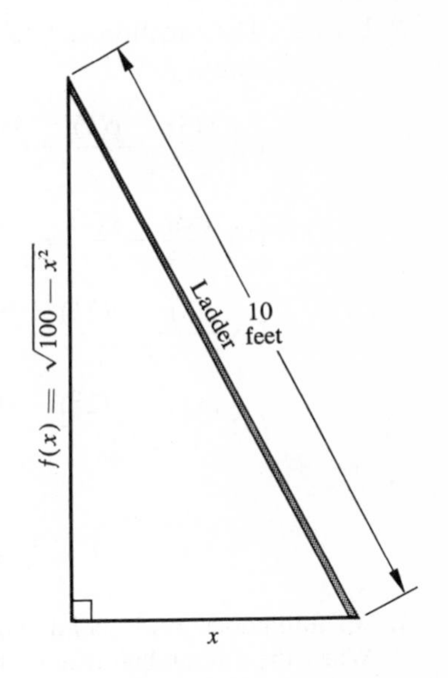

Figure 30

Solution: (a) $\dfrac{f(1) - f(0)}{1 - 0} = \sqrt{99} - 10 \approx -.05$

(b) $\dfrac{f(2) - f(1)}{2 - 1} = \sqrt{96} - \sqrt{99} \approx -.15$

(cont.)

(c) $\dfrac{f(3) - f(2)}{3 - 2} = \sqrt{91} - \sqrt{96} \approx -.26$

(d) $\dfrac{f(8) - f(6)}{8 - 6} = \dfrac{6 - 8}{2} = -1$

(e) $\dfrac{f(10) - f(8)}{10 - 8} = \dfrac{0 - 6}{2} = -3$

Example **(3)** Let g be a function such that $g(r)$ gives the area of a circle of radius r feet. Assume the radius r is being increased. Then the area, $g(r)$, is being increased accordingly. Find the average rate of change of the area as r is increased from

(a) 1 to 3 feet
(b) 2 to 3 feet
(c) $2\frac{1}{2}$ to 3 feet
(d) $2\frac{2}{3}$ to 3 feet

Solution: The function defined by $g(r) = \pi r^2$ gives the area for a circle of radius r.

(a) $\dfrac{g(3) - g(1)}{3 - 1} = \dfrac{9\pi - \pi}{2} = 4\pi$ sq. ft./foot

(b) $\dfrac{g(3) - g(2)}{3 - 2} = \dfrac{9\pi - 4\pi}{1} = 5\pi$ sq. ft./foot

(c) $\dfrac{g(3) - g(2\frac{1}{2})}{3 - 2\frac{1}{2}} = \dfrac{9\pi - \frac{25}{4}\pi}{\frac{1}{2}} = 5\frac{1}{2}\pi$ sq. ft./foot

(d) $\dfrac{g(3) - g(2\frac{2}{3})}{3 - 2\frac{2}{3}} = \dfrac{9\pi - \frac{64}{9}\pi}{\frac{1}{3}} = 5\frac{2}{3}\pi$ sq. ft./foot

EXERCISES

1. An automobile passes point A on a certain highway and moves toward point B. When the automobile reaches point B it stops, is immediately turned around, and returns along the same path to point A. At the end of t seconds the distance (in feet) of the automobile from point A is given by the function $f(t) = 100t - t^2$, $0 \leq t \leq 100$.

(a) Find the average speed of the automobile (in feet per second) in each of the following intervals:

$$[0, 10], [10, 20], [40, 50], [50, 60], [80, 90], [90, 100].$$

(b) Find the average speed of the automobile (in feet per second) in the interval [0, 50]. Convert this speed to miles per hour.

2. A man invested \$100.00 at 4 percent interest, compounded annually. The amount of the investment (principal plus interest) at the end of t years is given by the function $f(t) = 100(1.04)^t$. The values of $f(t)$ for $t = 1, 2, 3, \ldots, 20$ are listed.

(a) Find the average rate of change (in dollars and cents per year) for the following time intervals:

$$[0, 3], [3, 10], [0, 10], [10, 18], [18, 20].$$

(b) Approximate the average rate of change in each of the following intervals, using $f(18)$ to be exactly \$200.00:

$$[0, 36], [36, 72], [0, 72].$$

t	$f(t)$
1	104.00
2	108.16
3	112.49
4	116.99
5	121.67
6	126.53
7	131.59
8	136.86
9	142.33
10	148.02
11	153.95
12	160.10
13	166.51
14	173.17
15	180.09
16	187.30
17	194.79
18	202.58
19	210.68
20	219.11

3. Instantaneous Rate of Change

It is necessary to distinguish between *change* and *rate of change.* For example, "rate of change" may refer to the amount of change in relation to the amount of time elapsed. A change in distance of two miles by an automobile represents one rate if the time required is one hour and quite a different rate if the time required is only one minute; in the former case, the rate of change is an average of two miles per hour, whereas in the latter case the rate of change is an average of 120 miles per hour. It is not the *amount* of change that has taken place that may matter so much as it is the *rate* of change. As another example, a decrease in temperature of 80 degrees may cause no particular problems to a community if the change occurs over a period of several days or weeks. But if the change occurs overnight, severe damages may very likely result.

Consequently, the number $f(b) - f(a)$ is not so important in rate of change problems as is the number $\dfrac{f(b) - f(a)}{b - a}$, the average rate of change of $f(x)$ as x is increased from a to b.

Consider once again the example of the object which is dropped from a cliff 400 feet high and whose distance from the top of the cliff at any given time t is given by the function $f(t) = 16t^2$. This problem was discussed in the preceding section.

Given any interval $[a, b]$ which is a subset of the domain, $[0, 5]$, we can easily find the average rate of change of the falling object during that time interval by computing the difference quotient $\frac{f(b) - f(a)}{b - a}$. However, it is quite a different matter to define and determine the rate of change which is taking place at some precise instant. Let us describe a method for determining the rate of change taking place at a given instant, say at the time $t = 4$. To do so, we consider the following sequence of time intervals:

$$[4, 5], [4, 4\tfrac{1}{2}], [4, 4\tfrac{1}{3}], \cdots, \left[4, 4 + \frac{1}{n}\right], \cdots.$$

We form the corresponding sequence of numbers whose general term is

$$a_n = \frac{f\left(4 + \frac{1}{n}\right) - f(4)}{\frac{1}{n}} = \frac{16\left(4 + \frac{1}{n}\right)^2 - 16(4)^2}{\frac{1}{n}}$$

$$= 16n\left[\left(4 + \frac{1}{n}\right)^2 - 4^2\right] = 16n\left[4^2 + \frac{8}{n} + \frac{1}{n^2} - 4^2\right]$$

$$= 16n\left[\frac{8}{n} + \frac{1}{n^2}\right] = 128 + \frac{16}{n}.$$

The first eight terms of $\{a_n\}$ are 144, 136, $133\frac{1}{3}$, 132, $131\frac{1}{5}$, $130\frac{2}{3}$, $130\frac{2}{7}$ and 130. The limit of this sequence, 128, can be regarded as the *instantaneous rate of change* of the object at the precise instant when $t = 4$. We now have a very useful application of the theory of limits: The instantaneous rate of change of the falling object (indeed of *any* moving object) is a *limit!* Without the notion of a limit, this instantaneous rate of change could not be determined, for the ratio, rate $= \frac{\text{distance}}{\text{time}}$, is meaningless when the distance and the time are actually 0. The limit, 128, is the limit of a sequence of ratios of the form $\frac{D}{T}\left(\frac{\text{distance}}{\text{time}}\right)$, as D and T both approach 0 as a limit.

Example Each end section of a water trough is an isosceles triangle with base 4 feet and altitude 6 feet. The trough is 12 feet long. Water is pouring into the trough at the rate of 16 cubic feet per minute. (See Figure 31.)

(a) Find the depth $h(t)$ of the water t minutes after the water is turned on. How much time is required to fill the tank?

(b) Find the average rate of change of the depth of the water in each of the time intervals [0, 1], [1, 2], [2, 3], [3, 4], [4, 5], [5, 6], [6, 7], [7, 8], and [8, 9].

(c) Find the instantaneous rate of change of the depth of the water at the time $t = 4$ and also at the time $t = 9$.

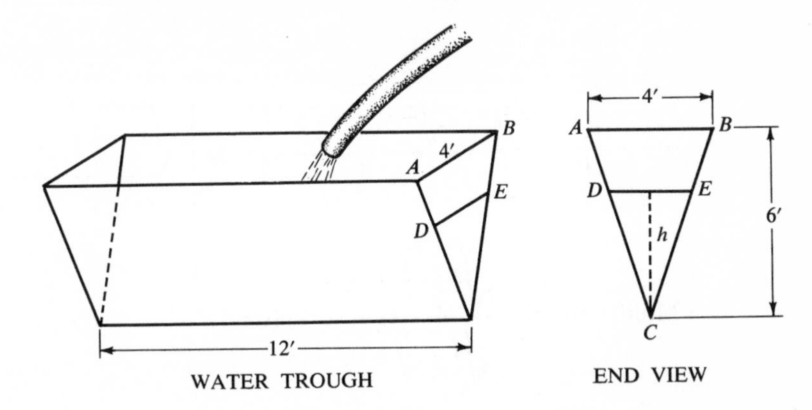

Figure 31

Solution: (a) Let DE represent the surface of the water at any time t. Let h be the altitude of ΔDEC to the base DE. We wish to find an expression for h in terms of t. Since triangles ABC and DEC are similar, the ratios DE/h and $\frac{4}{6}$ are equal, so that $DE = \frac{2}{3}h$ and the area of $\Delta DEC = \frac{1}{2}$(base)(altitude) $= \frac{1}{2}(\text{DE})(h) = \frac{1}{2}(\frac{2}{3}h)h = h^2/3$. The amount of water in the trough is then $(h^2/3)(12) = 4h^2$. Since water is flowing into the trough at the rate of 16 cubic feet per minute, we have $16t = 4h^2$ and $h = 2\sqrt{t}$. If we now define a function by $h(t) = 2\sqrt{t}$, this function will give the height of the water at time t. The tank is completely filled when $h(t) = 6$, and at this time $2\sqrt{t} = 6$ and $t = 9$. Thus a period of 9 minutes is required to fill the tank.

(b)

t	$h(t) = 2\sqrt{t}$ (to 2 decimal places)	Approximate Change in $h(t)$
1	2.00	2.00
2	$2\sqrt{2} = 2.82$	0.82
3	$2\sqrt{3} = 3.46$	0.64
4	4.00	0.54
5	$2\sqrt{5} = 4.47$	0.47
6	$2\sqrt{6} = 4.90$	0.43
7	$2\sqrt{7} = 5.30$	0.40
8	$2\sqrt{8} = 5.66$	0.36
9	6.00	0.34

The numbers in the right-hand column represent the rate of change of $h(t)$ in the unit intervals [0, 1], [1, 2], [2, 3], etc.

(c) Define a sequence by $a_n = \dfrac{h\left(4 + \frac{1}{n}\right) - h(4)}{\frac{1}{n}}$. Then

$$\frac{h\left(4+\frac{1}{n}\right) - h(4)}{\frac{1}{n}} = \frac{2\sqrt{4+\frac{1}{n}} - 2\sqrt{4}}{\frac{1}{n}} = 2n\left[\frac{\sqrt{4+\frac{1}{n}} - \sqrt{4}}{1}\right]$$

Multiplying by $\dfrac{\sqrt{4+\frac{1}{n}} + \sqrt{4}}{\sqrt{4+\frac{1}{n}} + \sqrt{4}}$ gives

$$2n\left[\frac{\sqrt{4+\frac{1}{n}} - \sqrt{4}}{1} \cdot \frac{\sqrt{4+\frac{1}{n}} + \sqrt{4}}{\sqrt{4+\frac{1}{n}} + \sqrt{4}}\right] = 2n\left[\frac{\left(4+\frac{1}{n}\right) - (4)}{\sqrt{4+\frac{1}{n}} + \sqrt{4}}\right]$$

$$= \frac{(2n)\left(\frac{1}{n}\right)}{\sqrt{4+\frac{1}{n}} + \sqrt{4}} = \frac{2}{\sqrt{4+\frac{1}{n}} + \sqrt{4}}.$$

The sequence $\left\{\sqrt{4+\frac{1}{n}}\right\} \to 2$, so from past experience we would conclude that $\{a_n\} = \left\{\dfrac{2}{\sqrt{4+\frac{1}{n}} + \sqrt{4}}\right\} \to \dfrac{2}{2+2} = \frac{1}{2}$.

(We will be able to prove this in Chapter 5.)
Thus the instantaneous rate of change of $h(t)$ at the time $t = 4$ is $\frac{1}{2}$ feet per minute.

Now let us find the instantaneous rate of change of $h(t)$ at the time $t = 9$. Define a sequence by

$$b_n = \frac{f\left(9 - \frac{1}{n}\right) - f(9)}{-\frac{1}{n}}.$$

Then

$$b_n = \frac{2\sqrt{9 - \frac{1}{n}} - 2\sqrt{9}}{-\frac{1}{n}}$$

$$= -2n\left[\frac{\sqrt{9 - \frac{1}{n}} - \sqrt{9}}{1} \cdot \frac{\sqrt{9 - \frac{1}{n}} + \sqrt{9}}{\sqrt{9 - \frac{1}{n}} + \sqrt{9}}\right]$$

$$= -2n\left[\frac{\left(9 - \frac{1}{n}\right) - 9}{\sqrt{9 - \frac{1}{n}} + \sqrt{9}}\right] = \frac{-2n\left(-\frac{1}{n}\right)}{\sqrt{9 - \frac{1}{n}} + \sqrt{9}}$$

$$= \frac{2}{\sqrt{9 - \frac{1}{n}} + \sqrt{9}}.$$

We conclude that $\{b_n\} \to \frac{2}{3 + 3} = \frac{1}{3}$.

REMARK: Since the function in the preceding example is not defined for $t > 9$, we are using $9 - \frac{1}{n}$ instead of $9 + \frac{1}{n}$. That is, instead of using the sequence $\{a_n\} = \left\{\frac{1}{n}\right\}$, we are using $\{b_n\} = \left\{-\frac{1}{n}\right\}$ to form the difference quotient. We could also have used $\{c_n\} = \left\{-\frac{1}{n^2}\right\}$ or any other sequence of negative numbers converging to 0.

In general, in forming difference quotients, we can use any sequence whose limit is 0, not merely $\{a_n\} = \left\{\frac{1}{n}\right\}$. For convenience, however, we ordinarily choose $\{a_n\} = \left\{\frac{1}{n}\right\}$. In subsequent exercises, you may use any sequence you wish which converges to 0, unless otherwise directed or unless you are limited to sequences of positive terms or sequences of negative terms, as in the preceding example.

Example *Rate of Change of the Quadratic Function f: $f(x) = x^2 - 4x$*

When x is in the interval	[0, 1]	[1, 2]	[2, 3]	[3, 4]	[4, 5]
The average rate of change is	−3	−1	1	3	5

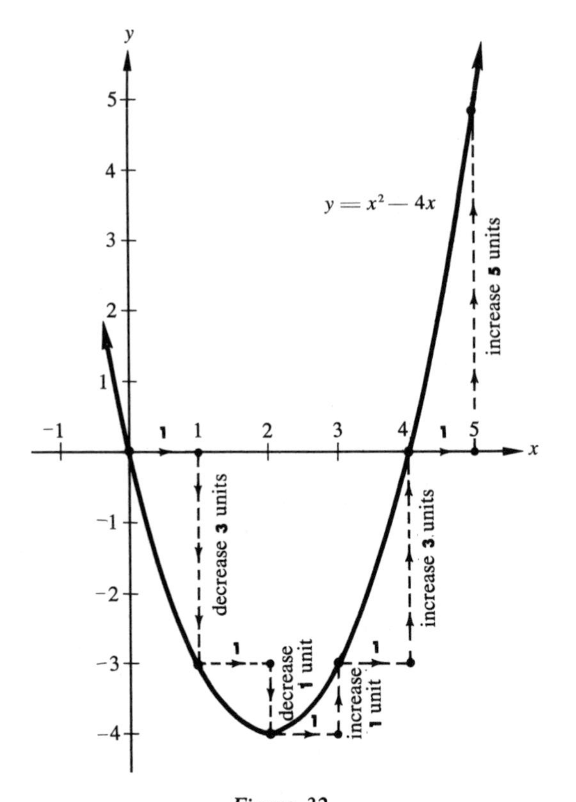

Figure 32

From the graph in Figure 32 the following observations can be made:

(1) As x is increased 1 unit from 0 to 1, $f(x)$ decreases 3 units. As x is increased 1 unit from 1 to 2, $f(x)$ decreases 1 unit. As x is increased 1 unit from 2 to 3, $f(x)$ increases 1 unit. As x is increased 1 unit from 3 to 4, $f(x)$ increases 3 units, etc.

(2) When x is in any interval $[a, b]$, where a and b are both less than 2, then the average rate of change of f is negative. Then for any number $c < 2$, the instantaneous rate of change at $x = c$,

which is the limit of a sequence of negative numbers, would appear to be negative.

(3) When x is in any interval $[a, b]$, where a and b are both greater than 2, then the average rate of change of f is positive. Thus for any number $c > 2$, the instantaneous rate of change would appear to be positive.

Let us now direct our attention to determining the rate at which $f(x)$ is changing at the precise instant when $x = 3$. Consider a sequence of intervals $[3, 4]$, $[3, 3\frac{1}{2}]$, $[3, 3\frac{1}{3}]$, $\cdots$, $\left[3, 3 + \frac{1}{n}\right]$, $\ldots$. The corresponding difference quotients are 3, $2\frac{1}{2}$, $2\frac{1}{3}$, $\cdots$ $2 + \frac{1}{n}$, $\cdots$, since

$$\frac{f\left(3 + \frac{1}{n}\right) - f(3)}{\frac{1}{n}} = \frac{\left(3 + \frac{1}{n}\right)^2 - 4\left(3 + \frac{1}{n}\right) - (-3)}{\frac{1}{n}}$$

$$= \frac{9 + \frac{6}{n} + \frac{1}{n^2} - 12 - \frac{4}{n} + 3}{\frac{1}{n}} = 2 + \frac{1}{n}.$$

The sequence $\{s_n\} = \left\{2 + \frac{1}{n}\right\}$ converges to 2, so 2 is the instantaneous rate of change of f at the precise moment when $x = 3$.

We are now ready to state a definition of instantaneous rate of change.

DEFINITION

Let f be a function and c a number in the domain of f. If, for all sequences $\{a_n\}$ which converge to 0, the sequence $\{b_n\} = \left\{\frac{f(c + a_n) - f(c)}{a_n}\right\}$ is well-defined and converges to a number R, then R is called the *instantaneous rate of change* of f at c. (See the following remarks.)

REMARKS: (1) The examples and exercises in this book have been selected so that for *any* choice of $\{a_n\}$ converging to 0, the resulting sequence $\{b_n\} = \left\{\frac{f(c + a_n) - f(c)}{a_n}\right\}$ converges if and only if the instantaneous rate of change does exist at c.

However, to apply the preceding definition to certain other functions, one must proceed much more generally and consider *all* possible sequences converging to 0. For example, if $f(x) = |x|$, at $x = 0$ the instantaneous rate of change does not exist; choosing

$$\{a_n\} = \left\{\frac{1}{n}\right\} \text{ gives } \{b_n\} = \left\{\frac{f\left(0 + \frac{1}{n}\right) - f(c)}{1/n}\right\} = \{1\} \longrightarrow 1$$

while choosing $\{a_n\} = \{-1/n\}$ gives $\{b_n\} = \{-1\} \longrightarrow -1$.

(2) If the function f in the preceding definition describes the position at any particular instant of an object moving in a straight line, then the phrase *instantaneous velocity* may be used instead of the phrase *instantaneous rate of change*, Accordingly, *instantaneous speed* (or simply speed) is defined as the absolute value of the instantaneous velocity.

Example Let $f(x) = x^3 + 2x$, $c = 3$ and $a_n = 1/n$.

Then $$b_n = \frac{\left[\left(3 + \frac{1}{n}\right)^3 + 2\left(3 + \frac{1}{n}\right)\right] - [3^3 + 2 \cdot 3]}{1/n}$$

(Recall the identity $(a + b)^3 = a^3 + 3a^2b + 3ab^2 + b^3$.)

$$= \frac{3^3 + \frac{27}{n} + \frac{9}{n^2} + \frac{1}{n^3} + 6 + \frac{2}{n} - 3^3 - 6}{1/n}$$

$$= n\left[\frac{27}{n} + \frac{9}{n^2} + \frac{1}{n^3} + \frac{2}{n}\right] = 29 + \frac{9}{n} + \frac{1}{n^2}.$$

From previous experience with sequences we conclude that $\{b_n\} \rightarrow 29$, so the instantaneous rate of change of f at $x = 3$ is 29. If x is in minutes and $f(x)$ is in feet, then this answer would be stated as 29 ft./min., or 29 feet per minute.

EXERCISES

1. A ball is rolled up an inclined plane. After it reaches the top of the plane, it returns to its starting point at the bottom. The distance, in feet, of the ball from the bottom of the plane is $24t - 4t^2 = f(t)$, where t is in seconds.
(a) Find the distance of the ball from the starting point at the end of 1 sec., 2 secs., 3 secs., 4 secs., 5 secs., and 6 secs. (b) Find the average velocity of the ball in each time interval:

$$[0, 1], [1, 2], [2, 3], [3, 4], [4, 5], \text{ and } [5, 6]$$

(c) Determine the instantaneous velocity of the ball at the precise instant $t = 2$ by finding the limit of the sequence given by $a_n = \dfrac{f\left(2 + \frac{1}{n}\right) - f(2)}{\frac{1}{n}}$.

(d) Determine how fast the ball is rolling when it reaches the bottom of the inclined plane on its return trip.

REMARK: *Speed of an object* moving in a straight line is defined as the *absolute value of the velocity.* In Problem 1 above, the average speed in the interval [2, 3] is the same as the average speed in the interval [3, 4]. Only the direction of the moving object has changed. In this problem *positive velocity* means that as t is increased, $f(t)$, the distance from the ball to the starting point, is increased. *Negative velocity* means that as t is increased, $f(t)$ is decreased.

2. A farm lighting system has a generator driven by a gasoline motor. The number of gallons of gasoline burned is given by the formula $g(t) = \frac{2}{7}t$, where t is the number of hours the motor has been running.

(a) Determine the number of gallons of gasoline burned in each time interval: [0, 1], [1, 2], [2, 3], [3, 4].

(b) Find the instantaneous rate of change of g at any time t.

(c) Find the instantaneous rate of change of g at the time when $t = 3$.

3. Let A (the area of a square with side x) be given by $f(x) = x^2$.

(a) How much is A increased when x is increased from 0 to 1? 1 to 2? 2 to 3? 3 to 4? What is the average rate of change of A in the interval [0, 1]? in [1, 2]? in [2, 3]? in [3, 4]?

(b) Simplify the expression

$$a_n = \frac{f\left(3 + \frac{1}{n}\right) - f(3)}{\frac{1}{n}}$$ and then find the limit, if it exists, of $\{a_n\}$.

(c) Consider x to be any real number. Simplify the expression

$$b_n = \frac{f\left(x + \frac{1}{n}\right) - f(x)}{\frac{1}{n}}$$ and find the limit of $\{b_n\}$.

(Your answer will be in terms of x.) What is this limit when $x = 2$? when $x = 3$? when $x = 4$?

4. Three grams of a certain chemical compound are dissolved in a solution. The amount of the compound in the solution decreases with time, so that $A = 3/t$ gives the number of grams remaining after t hours.

(a) What is the *average* rate of loss per hour from the end of 1 hour to the end of 4 hours?

(b) What is the *average* rate of loss per hour from the end of 4 hours to the end of $4\frac{1}{2}$ hours? from the end of 4 hours to the end of $4\frac{1}{3}$ hours?

(c) Define a sequence whose limit can be regarded as the *instantaneous* rate of loss of the compound at the time $t = 4$. Find this limit.

(d) Define a sequence whose limit can be regarded as the *instantaneous* rate of loss of the compound at *any* time t. Find the limit of this sequence (in terms of t).

5. A projectile strikes an earthen bank and penetrates a certain distance. After t seconds the projectile has traveled $y = f(t) = 80t - 8t^2$ inches, $0 \leq t \leq 5$. Assume that t, and hence y, is 0 at the time of impact. Assume that the velocity of the projectile at any time t is given by the formula $v(t) = 80 - 16t$ inches per second.

(a) How far will the projectile have penetrated when it comes to a stop?

(b) What is the velocity of the projectile at the time $t = 0$? at $t = 3$? at $t = 5$?

(c) Show why the formula for the velocity at any time is $80 - 16t$.

6. Let $g(x) = y = x^2 - x$.

(a) Graph this function in the coordinate plane.

(b) Find the instantaneous rate of change of y with respect to x at $x = 1$, at $x = \frac{1}{2}$, and at $x = 0$.

4. Tangent to a Curve

For a given function f and a number c in the domain of f, we have determined what it means to speak of the *instantaneous rate of change* of f at c. Let us now proceed further and investigate the notion of the tangent line to the graph of f at the point $(c, f(c))$. These two ideas, as we shall notice, are directly related.

For example, let $f(x) = x^2 + 1$ and $c = 2$. For larger and larger values of n, the value of $2 + \dfrac{1}{n}$ would come closer and closer to 2, and the value of

$$\frac{f\left(2 + \frac{1}{n}\right) - f(2)}{\frac{1}{n}} = \frac{\left[\left(2 + \frac{1}{n}\right)^2 + 1\right] - [2^2 + 1]}{\frac{1}{n}} = 4 + \frac{1}{n}$$

would come closer and closer to the number 4. *This number, 4, represents the instantaneous rate of change of the function at the precise moment when*

$x = 2$. Now let us determine what this number, 4, means as far as the graph of the function, $f(x) = x^2 + 1$, is concerned.

In the ensuing discussion, a vivid imagination will be helpful, for we wish to imagine that the graph of $f(x) = x^2 + 1$ has been enlarged greatly. (See Figure 33.) In order to facilitate illustration, we have drawn a pair of coordinate axes with origin at (1, 4), rather than the regular x, y coordinate axes with origin at (0, 0).

Let A be the point $(2, f(2)) = (2, 5)$ on the graph, and consider a sequence of points $\{B_n\}$ on the graph, where

$$B_n = \left(2 + \frac{1}{n}, f\left(2 + \frac{1}{n}\right)\right)$$
$$= \left(2 + \frac{1}{n}, \left(2 + \frac{1}{n}\right)^2 + 1\right)$$
$$= \left(2 + \frac{1}{n}, 5 + \frac{4}{n} + \frac{1}{n^2}\right).$$

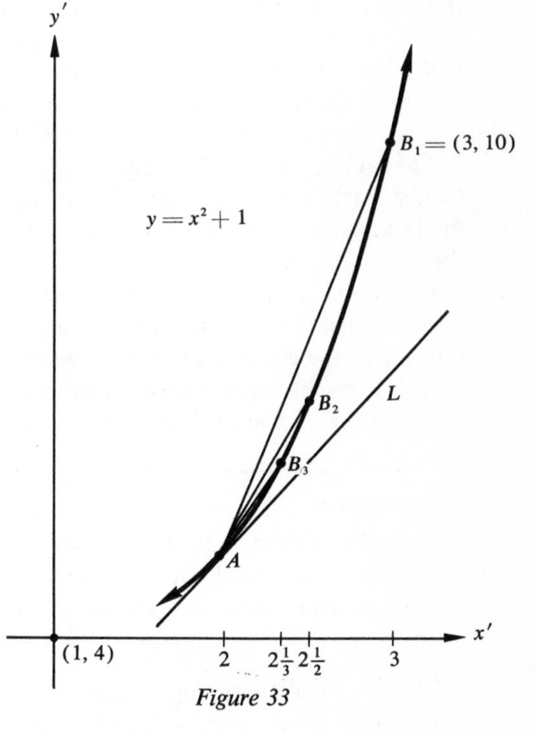

Figure 33

Then

$B_1 = (3, 10)$
$B_2 = (2\frac{1}{2}, 7\frac{1}{4})$
$B_3 = (2\frac{1}{3}, 6\frac{4}{9})$
$B_4 = (2\frac{1}{4}, 6\frac{1}{16})$
$\cdot \quad \cdot$
$\cdot \quad \cdot$
$\cdot \quad \cdot$

$$B_n = \left(2 + \frac{1}{n}, 5 + \frac{4}{n} + \frac{1}{n^2}\right).$$

Choose a very large natural number, say 1,000,000. The arc of the curve from A to $B_{1,000,000}$ is virtually indistinguishable from the line segment from A to $B_{1,000,000}$, and the slope of this line segment is
$$\frac{f\left(2 + \dfrac{1}{1,000,000}\right) - f(2)}{\dfrac{1}{1,000,000}} = 4 + \frac{1}{1,000,000}$$
$= 4.000001$.

It seems reasonable to expect that for larger and larger natural numbers n the arcs from A to B_n will become less and less distinguishable from their corresponding line segments from A to B_n. Furthermore, the *slopes* of these line segments are the terms of the sequence

$$\{a_n\} = \left\{ \frac{f\left(2 + \frac{1}{n}\right) - f(2)}{\frac{1}{n}} \right\} = \left\{4 + \frac{1}{n}\right\}.$$

For larger and larger natural numbers n, these line segments will come closer and closer to the line which passes through A, with slope of exactly 4. (Call this line L.) In fact, we can say that L is the limit of the sequence of line segments. Line L is referred to as the *tangent* to the curve at point A. Accordingly, the slope of line L is the same number as the instantaneous rate of change of the function at the precise moment when $x = 2$! In fact, the slope of L really represents what we could call the *direction of the curve at point A*.

The same effect can be realized by imagining point A to be a fixed point on the Earth's equator. Let $\{B_n\}$ be a sequence of points lying west of A on the equator, each point being an arc distance of $1/n$ miles from A. Suppose the points B_n represent the various locations of a small ship sailing east toward point A. For larger and larger values of n, the arc from B_n to A will *almost* coincide with a straight line segment connecting the two points. The tangent to the equator at point A would seem to lie *flat* on the ocean and would represent the horizontal direction at point A.

It now seems appropriate to state a definition of tangent to a curve.

DEFINITION

Let f be a function, $(c, f(c))$ a point on the graph of f, and R the instantaneous rate of change of f at c, if it exists. The line which passes through $(c, f(c))$ and has slope R is called the *tangent to the graph of f at the point* $(c, f(c))$.

The matter of specifying what is meant by a *tangent to a curve at a given point* on the curve is not a difficult task, using the language of limits. However, without the precise language of limits, the task is not a simple one. In your courses in plane geometry the notion of a tangent to a circle was not at all difficult to define. For a given point P on the circle, the tangent to the circle at P was probably defined to be the line which passes through point P and only point P on the circle. However, for curves which are not circles, this definition would not suffice, as the curve in Figure 34 illustrates.

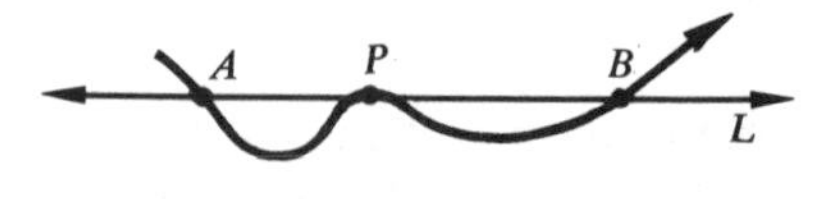

Figure 34

In this figure, line L, which is tangent to the curve at point P, passes through points of the curve other than P, namely A and B. Moreover, there are infinitely many lines which are not tangent to the curve at P and yet which pass through P and only point P of the curve.

Example Find the equation of the line which is tangent to the graph of $f(x) = x^3 - x$ at the point $(2, f(2))$.

Solution: Form the sequence

$$\{b_n\} = \left\{\frac{f\left(2+\frac{1}{n}\right) - f(2)}{\frac{1}{n}}\right\} = \left\{\frac{\left[\left(2+\frac{1}{n}\right)^3 - \left(2+\frac{1}{n}\right)\right] - [2^3 - 2]}{\frac{1}{n}}\right\}$$

$$= \left\{\frac{2^3 + 3\cdot 2^2\cdot\frac{1}{n} + 3\cdot 2\cdot\frac{1}{n^2} + \frac{1}{n^3} - 2 - \frac{1}{n} - 2^3 + 2}{\frac{1}{n}}\right\}$$

$$= \left\{n\left[\frac{12}{n} + \frac{6}{n^2} + \frac{1}{n^3} - \frac{1}{n}\right]\right\} = \left\{11 + \frac{6}{n} + \frac{1}{n^2}\right\}.$$

Then $\{b_n\} \to 11$.

Thus the slope of the tangent line at the point $(2, 6)$ is 11.

Using the equation $y = mx + d$, we obtain $6 = 11(2) + d$ and $d = 6 - 22 = -16$. This means that the equation for the tangent line at point $(2, 6)$ is $y = 11x - 16$. (See Figure 35.)

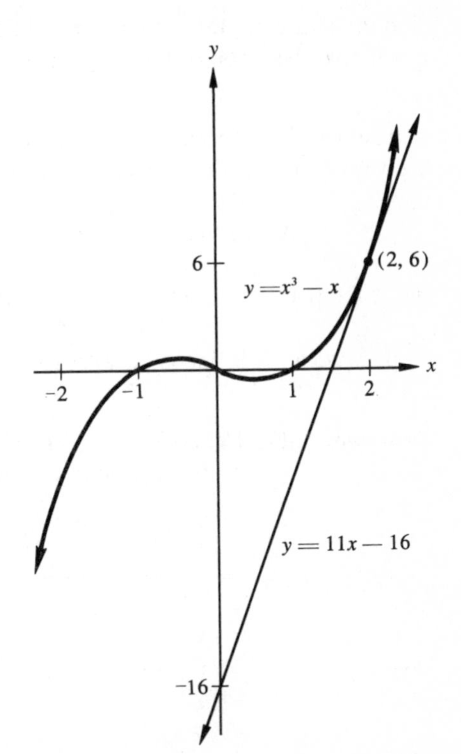

Figure 35

EXERCISES

1. Let $f(x) = x^3$. Find the equation of the tangent line at
   (a) point $(1, 1)$; (b) point $(-1, -1)$; (c) point $(0, 0)$.
   Include a graph of the function, showing these three tangent lines.
2. Find the equation of the line which is tangent to the graph of $f(x) = x^3 - x$ at (a) point $(1, 0)$; (b) point $(0, 0)$. Include a graph of the function, showing these two tangent lines.
3. Find the equation of the line which is tangent to the graph of $f(x) = x^2 - 2x + 1$ at (a) point $(2, 1)$; (b) point $(3, 4)$; (c) point $(0, 1)$; (d) point $(1, 0)$. Include a graph of the function, showing these four tangent lines.

5. Further Discussion of Tangent Lines

Let $(c, f(c))$ be any point on the graph of a function. If the limit of the sequence $\{a_n\} = \left\{\dfrac{f\left(c + \frac{1}{n}\right) - f(c)}{\frac{1}{n}}\right\}$ exists, then this limit will be an expression involving c, and evaluation of this expression for any specified value of c will give the slope of the tangent to the curve at the point $(c, f(c))$.

REMARK: You should keep in mind that a line in the coordinate plane has a positive slope if the line slants upward as we examine it from left to right: ╱. A line has negative slope if it slants downward as we examine it from left to right: ╲. If a line is parallel to the x-axis, its slope is zero. If a line is parallel to the y-axis, its slope is undefined.

Example **(1)** Let $f(x) = x^2 + 1$, with domain the set of all real numbers. If c is any real number, then

$$\{a_n\} = \left\{\frac{f\left(c + \frac{1}{n}\right) - f(c)}{\frac{1}{n}}\right\} = \left\{\frac{\left[\left(c + \frac{1}{n}\right)^2 + 1\right] - [c^2 + 1]}{\frac{1}{n}}\right\}$$

$$= \left\{2c + \frac{1}{n}\right\} \to 2c.$$

In particular, for $c = -3, -2, -1, 0, 1, 2$, and 3, the slopes of the corresponding tangent lines are $-6, -4, -2, 0, 2, 4$, and 6, respectively. (See Figure 36a.) The equations of these tangents are:

$y = -6x - 8$ $\qquad$ $y = 2x$
$y = -4x - 3$ $\qquad$ $y = 4x - 3$
$y = -2x$ $\qquad$ $y = 6x - 8$
$y = 0 \cdot x + 1$ (or simply $y = 1$)

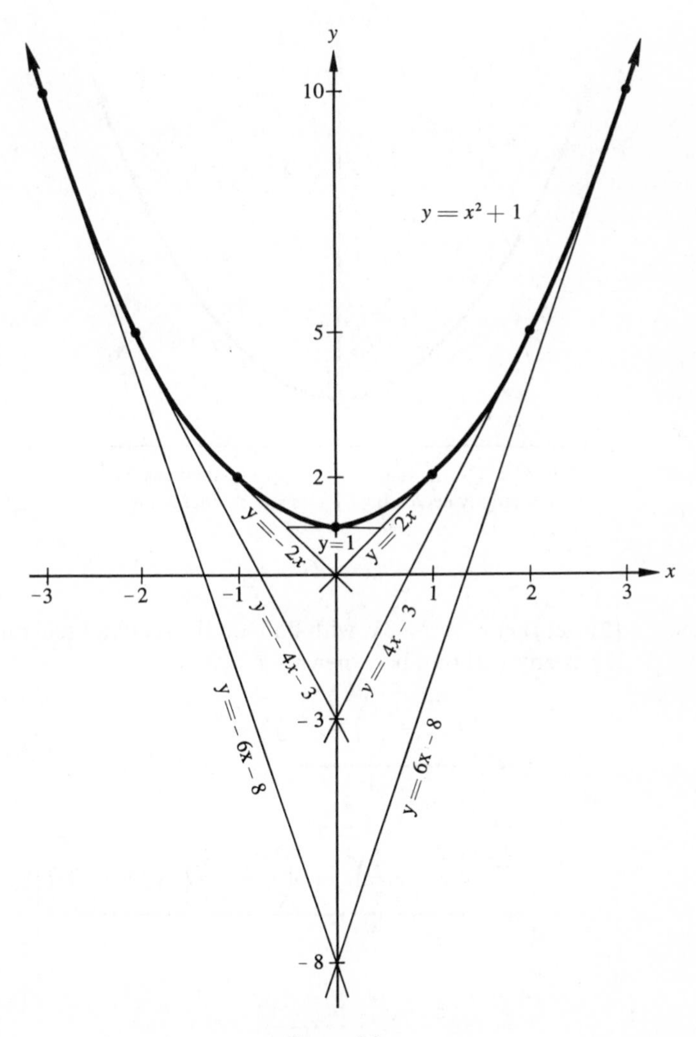

Figure 36a

Notice that $2c < 0 \leftrightarrow c < 0$, so that for these values of c the slope of the tangent line at $(c, f(c))$ is negative, and that in the interval $\langle -\infty, 0\rangle$ the function is strictly decreasing. (See Figure 36b.)

Since $2c > 0 \leftrightarrow c > 0$, for these values of c the slope of the tangent line at $(c, f(c))$ is positive. Notice that in the interval $\langle 0, \infty\rangle$ the function is strictly increasing.

Since $2c = 0 \leftrightarrow c = 0$, at the point $(0, 1)$ the tangent line has slope of 0 and is therefore horizontal.

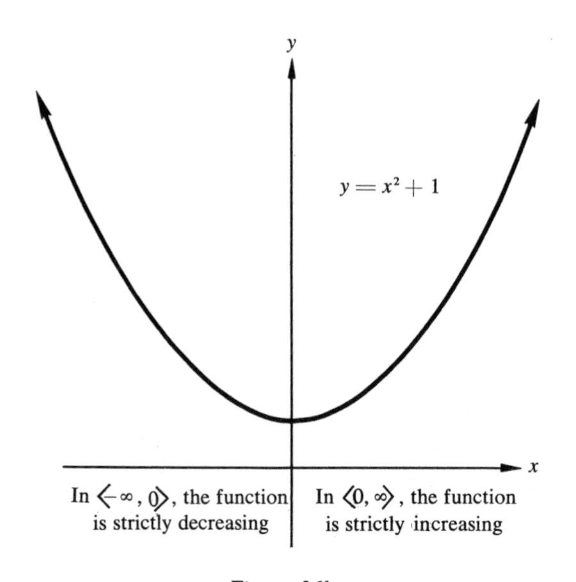

Figure 36b

Example **(2)** Let $f(x) = x^3 - 3x$, with domain the set of all real numbers. If c is any real number, then

$$\{b_n\} = \left\{\frac{f\left(c + \frac{1}{n}\right) - f(c)}{\frac{1}{n}}\right\}$$

$$= \left\{\frac{\left[\left(c + \frac{1}{n}\right)^3 - 3\left(c + \frac{1}{n}\right)\right] - [c^3 - 3c]}{\frac{1}{n}}\right\}$$

$$= \left\{3c^2 - 3 + \frac{3c}{n} + \frac{1}{n^2}\right\} \to 3c^2 - 3.$$

Thus at any point $(c, f(c))$ on the graph of f the slope of the tangent line is $3c^2 - 3$. Notice that $3c^2 - 3 < 0 \leftrightarrow c^2 < 1 \leftrightarrow -1 < c < 1$. For these values of c the slope of the tangent line at $(c, f(c))$ is negative, and the function is strictly decreasing in the interval $\langle -1, 1 \rangle$.

Also, $3c^2 - 3 > 0 \leftrightarrow c^2 > 1 \leftrightarrow (c < -1 \text{ or } c > 1)$. For these values the slope of the tangent line at $(c, f(c))$ is positive, and the function is strictly increasing in the intervals $\langle -\infty, -1 \rangle$ and $\langle 1, \infty \rangle$.

Since $3c^2 - 3 = 0 \leftrightarrow c = -1$ or $c = 1$, then at the points $(-1, 2)$ and $(1, -2)$ the tangent line has a slope of 0 and is therefore horizontal. (See Figure 37.)

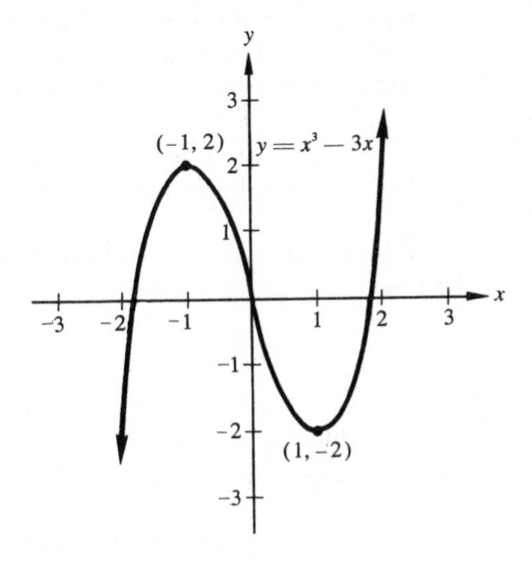

Figure 37

Let us state the following definitions and then discuss their applications to the preceding examples.

DEFINITIONS

(1) A function f is said to have a *relative minimum* at a number c in the domain of f if there is some neighborhood of c such that, for all numbers x (except c) which are in the domain of f and in this neighborhood, $f(x) > f(c)$.

(2) A function f is said to have a *relative maximum* at a number c in the domain of f if there is some neighborhood of c such that, for all numbers x (except c) which are in the domain of f and in this neighborhood, $f(x) < f(c)$.

The function whose graph appears in Figure 36b attains a relative minimum at $x = 0$. The function whose graph appears in Figure 37 attains a relative maximum at $x = -1$ and a relative minimum at $x = 1$. The following remarks relate to these two functions:

REMARKS: (1) At each of the three points mentioned, the slope of the tangent line is 0. In order for a function to have a relative maximum or a relative minimum at a number c in its domain, it is *necessary* that the slope of the tangent line at $(c, f(c))$ be 0, but it is *not sufficient*. This means that it is possible for the slope of the tangent line at point $(c, f(c))$ to be 0 without the function having either a relative maximum or a relative minimum at c, as a later example will demonstrate (Example (3), opposite).

(2) In Figure 36b, we observe that just to the left of $x = 0$ the slopes of all tangent lines to the graph of the function would be negative, whereas just to the right of $x = 0$ the slopes of all tangent lines to the graph of the function would be positive.

(3) In Figure 37, just to the left of $x = 1$ the slopes of the tangent lines to the graph would be negative, whereas just to the right of $x = 1$ they would be positive.

(4) In Figure 37, just to the left of $x = -1$ the slopes of the tangent lines to the graph would be positive, whereas just to the right of $x = -1$ they would be negative.

Remarks 1, 2, and 3 above lead us to the following generalization:

Let f be a function and c a number in the domain of f such that at point $(c, f(c))$ the tangent line exists and has a slope of 0. Suppose further that there is a neighborhood $\langle a,b \rangle$ of c such that, whenever x is in $\langle a,c \rangle$, the slope of the tangent line at $(x, f(x))$ is negative, *and*, whenever x is in $\langle c,b \rangle$, the slope of the tangent line at $(x, f(x))$ is positive. Then, under these circumstances, f has a relative minimum at $x = c$.

Remarks 1 and 4 above lead us to the following generalization:

Let f be a function and c a number in the domain of f such that at point $(c, f(c))$ the tangent line exists and has a slope of 0. Suppose further that there is a neighborhood $\langle a,b \rangle$ of c such that, whenever x is in $\langle a,c \rangle$, the slope of the tangent line at $(x, f(x))$ is positive, *and*, whenever x is in $\langle c,b \rangle$, the slope of the tangent line at $(x, f(x))$ is negative. Then, under these circumstances, f has a relative maximum at $x = c$.

In our work with functions and tangents, we may have already made the following important and useful observation: The slopes of the tangent lines to the graph of a function over a given open interval $\langle a, b \rangle$ are positive if and only if the function is strictly increasing on the interval $\langle a, b \rangle$; they are negative if and only if the function is strictly decreasing on $\langle a, b \rangle$.

Using the above observation, we can now make this further generalization: Let f be a function and c a number in the domain of f such that at point $(c, f(c))$ the tangent line exists and has a slope of 0. Then

(i) if the function is strictly decreasing on some open interval $\langle b,c \rangle$ with $b < c$ and is strictly increasing on some open interval $\langle c,d \rangle$ with $c < d$, then the function has a relative minimum at c.

(ii) if the function is strictly increasing on some open interval $\langle b,c \rangle$ with $b < c$ and is strictly decreasing on some open interval $\langle c,d \rangle$ with $c < d$, then the function has a relative maximum at c.

The following example illustrates a point made in Remark (1) on the preceding page.

Example **(3)** Let $f(x) = x^3$, with domain the set of all real numbers. If c is any real number, then

$$\{d_n\} = \left\{\frac{f\left(c + \frac{1}{n}\right) - f(c)}{\frac{1}{n}}\right\}$$

$$= \left\{\frac{\left(c + \frac{1}{n}\right)^3 - c^3}{\frac{1}{n}}\right\}$$

$$= \left\{\frac{c^3 + \frac{3c^2}{n} + \frac{3c}{n^2} + \frac{1}{n^3} - c^3}{\frac{1}{n}}\right\}$$

$$= \left\{3c^2 + \frac{3c}{n} + \frac{1}{n^2}\right\} \to 3c^2.$$

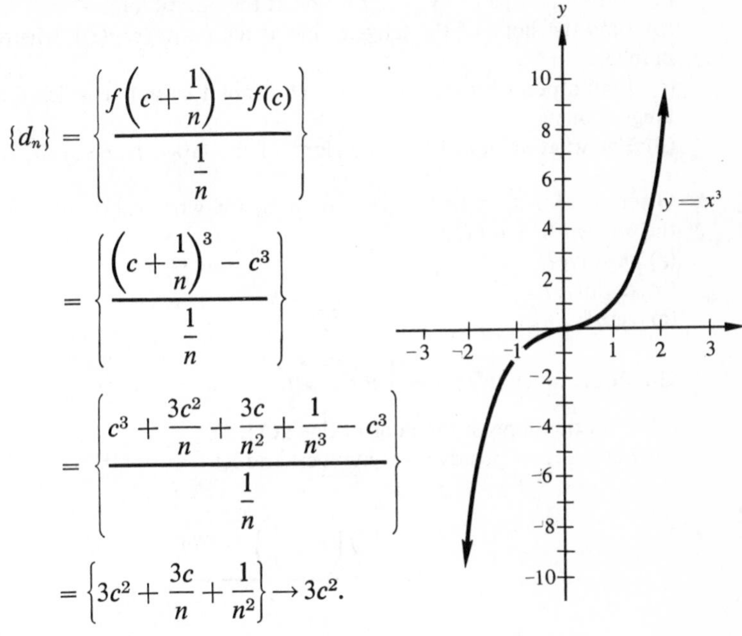

Figure 38

Since $3c^2 = 0 \leftrightarrow c = 0$, then at $(0, 0)$ the slope of the tangent line is 0. However, just to the left of 0 the function is strictly increasing, and just to the right of 0 the function is also strictly increasing, so the function attains neither a relative maximum nor a relative minimum at $x = 0$. (See Figure 38.)

The main purpose of this section has been to introduce the concepts of *relative minimum* and *relative maximum* of a function, together with methods for determining them for a particular function. These methods make use of the notions *strictly increasing* and *strictly decreasing* and of the closely related notions *positive slope* and *negative slope*.

In courses in calculus you will find that relative maxima and relative minima are very important tools for studying the behavior of functions. In these later courses more sophisticated methods for determining them will be introduced.

EXERCISES

1. Graph the function $g(x) = x^2 + x$.
(a) Find the slope of the tangent line at the point (2, 6).
(b) Find the slope of the tangent line at the point $(c, g(c))$, where c is any real number.
(c) Find a point on the curve (if such a point exists) at which the slope of the tangent line is 0.
(d) For what values of x is the slope of the tangent line at $(x, f(x))$ negative?

2. Given the function $f(x) = x^2 + 6x + 5$, for what values of x is the slope of the tangent line at $(x, f(x))$
(a) positive?
(b) negative?
(c) zero?

3. Graph the function $f(x) = \dfrac{1}{x}$ for $x > 0$.
(a) Find the slope of the tangent line at (1, 1).
(b) Let c be any positive real number, and let

$$s_n = \frac{f\left(c + \frac{1}{n}\right) - f(c)}{\frac{1}{n}}.$$

Show that s_n simplifies to $\dfrac{-1}{c\left(c + \frac{1}{n}\right)}$, and find the limit of $\{s_n\}$ for $c = 1, 2, 10$.

In general, $\{s_n\}$ converges to what number (in terms of c)?
(c) Using the results of part (b), find the equations of the tangent lines at points $(2, \frac{1}{2})$ and $(\frac{1}{3}, 3)$. Sketch these lines on your graph.

4. Consider a family of curves of the form $y = ax^2$, such that every real number a (except 0) determines a member of this family.
(a) If $a = 1$, then at any point (c, c^2) on the curve $y = x^2$, the slope of the tangent line is __?__.
(b) If $a = 2$, then at any point $(c, 2c^2)$ on the curve $y = 2x^2$, the slope of the tangent line is __?__.
(c) If a is unspecified, then at any point (c, ac^2) on the curve $y = ax^2$, the slope of the tangent line is __?__.

5. Graph the function $f(x) = \sin x$, $0 \leq x \leq 2\pi$. It is known that for any number c the slope of the tangent line at the point $(c, \sin c)$ is the number $\cos c$.
(a) For what numbers x is the slope of the tangent line 0?
(b) For what numbers x is the slope of the tangent line -1?

5 • *Operations with Sequences*

1. Sums, Differences, and Products

Numbers are not the only elements in mathematics which we can add, subtract, multiply and divide. The familiar operations of arithmetic can also be performed with functions (in particular, *sequences*), with vectors (directed line segments), and with many other kinds of elements, including matrices. (A matrix, as you may know, is an array of numbers arranged in rows and columns.) We will discuss operations with sequences in this chapter and will arrive at several theorems which will enable you to calculate limits quickly for a great many sequences.

DEFINITIONS

Given any two sequences $\{a_n\}$ and $\{b_n\}$, then

(1) the *sum* of $\{a_n\}$ and $\{b_n\}$ is defined to be the sequence whose *n*th term is $a_n + b_n$. In notation, $\{a_n\} + \{b_n\} = \{a_n + b_n\}$.

(2) the *difference* of $\{a_n\}$ and $\{b_n\}$ is defined to be the sequence whose *n*th term is $a_n - b_n$. Thus $\{a_n\} - \{b_n\} = \{a_n - b_n\}$.

(3) the *product* of $\{a_n\}$ and $\{b_n\}$ is defined to be the sequence whose *n*th term is $a_n b_n$. Thus $\{a_n\}\{b_n\} = \{a_n b_n\}$.

For example,

For the Sequence	*The First Nine Terms Are*								
$\{a_n\} = \left\{\frac{n+1}{n}\right\}$	2	$1\frac{1}{2}$	$1\frac{1}{3}$	$1\frac{1}{4}$	$1\frac{1}{5}$	$1\frac{1}{6}$	$1\frac{1}{7}$	$1\frac{1}{8}$	$1\frac{1}{9}$
$\{b_n\} = \left\{\frac{2n+3}{3n}\right\}$	$1\frac{2}{3}$	$1\frac{1}{6}$	1	$\frac{11}{12}$	$\frac{13}{15}$	$\frac{5}{6}$	$\frac{17}{21}$	$\frac{19}{24}$	$\frac{7}{9}$

Using these nine pairs of terms, we obtain the first nine terms of

$$\begin{array}{llllllllll}
\{a_n + b_n\}: & 3\frac{2}{3} & 2\frac{2}{3} & 2\frac{1}{3} & 2\frac{1}{6} & 2\frac{1}{15} & 2 & 1\frac{20}{21} & 1\frac{11}{12} & 1\frac{8}{9} \\
\{a_n - b_n\}: & \frac{1}{3} & \frac{1}{3} & \frac{1}{3} & \frac{1}{3} & \frac{1}{3} & \frac{1}{3} & \frac{1}{3} & \frac{1}{3} & \frac{1}{3} \\
\{a_n b_n\}: & 3\frac{1}{3} & 1\frac{3}{4} & 1\frac{1}{3} & 1\frac{7}{48} & 1\frac{1}{25} & \frac{35}{36} & \frac{136}{147} & \frac{57}{64} & \frac{70}{81}
\end{array}$$

The general terms for these three new sequences are

$$a_n + b_n = \frac{5n + 6}{3n}, \quad a_n - b_n = \tfrac{1}{3}, \quad \text{and} \quad a_n b_n = \frac{2n^2 + 5n + 3}{3n^2}.$$

Assuming that $\{a_n\} \to 1$ and $\{b_n\} \to \frac{2}{3}$, it would appear that $\{a_n + b_n\} \to 1 + \frac{2}{3} = 1\frac{2}{3}$, $\{a_n - b_n\} \to 1 - \frac{2}{3} = \frac{1}{3}$ and $\{a_n b_n\} \to 1 \cdot \frac{2}{3} = \frac{2}{3}$. These conclusions are true, as a result of the following three theorems. The validity of these theorems should not be surprising, in view of our work to date with sequences. Later in this chapter we will discuss proofs of these theorems.

THEOREM 5-1

If $\{a_n\} \to A$ and $\{b_n\} \to B$, then $\{a_n + b_n\} \to A + B$.

THEOREM 5-2

If $\{a_n\} \to A$ and $\{b_n\} \to B$, then $\{a_n - b_n\} \to A - B$.

THEOREM 5-3

If $\{a_n\} \to A$ and $\{b_n\} \to B$, then $\{a_n b_n\} \to AB$.

Applications of Theorems 5-1, 5-2, 5-3

In the following applications of Theorems 5-1, 5-2, and 5-3, it is helpful to know convergence for certain basic sequences. Let us therefore assume that any constant sequence $\{c\}$ converges to c, that $\{a_n\} = \left\{\frac{n}{n+1}\right\} \to 1$ (this will be proved in a later section of this chapter), and that $\{b_n\} = \left\{\frac{1}{n^p}\right\} \to 0$, where p is any positive real number. (For example, if $p = 1$, we have $\{c_n\} = \left\{\frac{1}{n}\right\} \to 0$, and if $p = 2$, we have $\{d_n\} = \left\{\frac{1}{n^2}\right\} \to 0$.) The student is asked to prove $\{b_n\} \to 0$ and $\{c\} \to c$ in the Exercises of this section.

Using these assumptions, we can proceed to the following applications.

(1) Let us find the limit of $\{a_n\} = \left\{\frac{5n+1}{n}\right\}$. Since $\left\{\frac{5n+1}{n}\right\} = \left\{\frac{5n}{n} + \frac{1}{n}\right\} = \{5\} + \left\{\frac{1}{n}\right\}$, then by Theorem 5-1 and the above assumptions, $\{a_n\} \to 5$.

In general, if a sequence $\{a_n\}$ is the sum of two sequences, one of which is a constant sequence — say, $\{a_n\} = \{c\} + \{b_n\}$ — then geometrically the points representing $\{a_n\}$ occupy the same position relative to each other as the points representing $\{b_n\}$. (See Figure 39.) In such an event, $\{a_n\}$ converges if and only if $\{b_n\}$ converges.

Thus we note that the sequences $\{c_n\} = \left\{\frac{2n+1}{n}\right\}$, $\{d_n\} = \left\{\frac{2n^2+1}{n^2}\right\}$, and $\{e_n\} = \left\{\frac{2n^3+1}{n^3}\right\}$ all converge to 2.

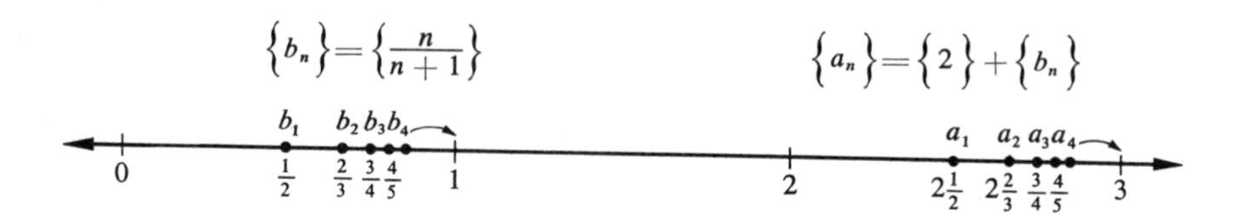

Figure 39

(2) Theorem 5-3 has many applications in the event one of the two sequences is a constant sequence:

$$\{a_n\} = \left\{\frac{2}{n}\right\} = \{2\}\left\{\frac{1}{n}\right\} \to 2 \cdot 0 = 0$$

$$\{b_n\} = \left\{\frac{5}{n^2}\right\} = \{5\}\left\{\frac{1}{n^2}\right\} \to 5 \cdot 0 = 0$$

$$\{c_n\} = \left\{\frac{4n}{n+1}\right\} = \{4\}\left\{\frac{n}{n+1}\right\} \to 4 \cdot 1 = 4$$

$$\{d_n\} = \left\{\frac{2n+3}{n}\right\} = \{2\} + \{3\}\left\{\frac{1}{n}\right\} \to 2 + 3 \cdot 0 = 2$$

$$\{e_n\} = \left\{\frac{n+3}{4n}\right\} = \left\{\frac{1}{4}\right\}\left\{\frac{n+3}{n}\right\} = \left\{\frac{1}{4}\right\}\left\{1 + (3)\left(\frac{1}{n}\right)\right\} \to \tfrac{1}{4}(1 + 3 \cdot 0) = \tfrac{1}{4}.$$

In Figure 40 on the next page, we are assuming that $\{a_n\} = \left\{\frac{n-1}{n+1}\right\} \to 1$, and we are multiplying this sequence by the constant sequences $\{2\}$, $\{\frac{1}{2}\}$, and $\{-2\}$.

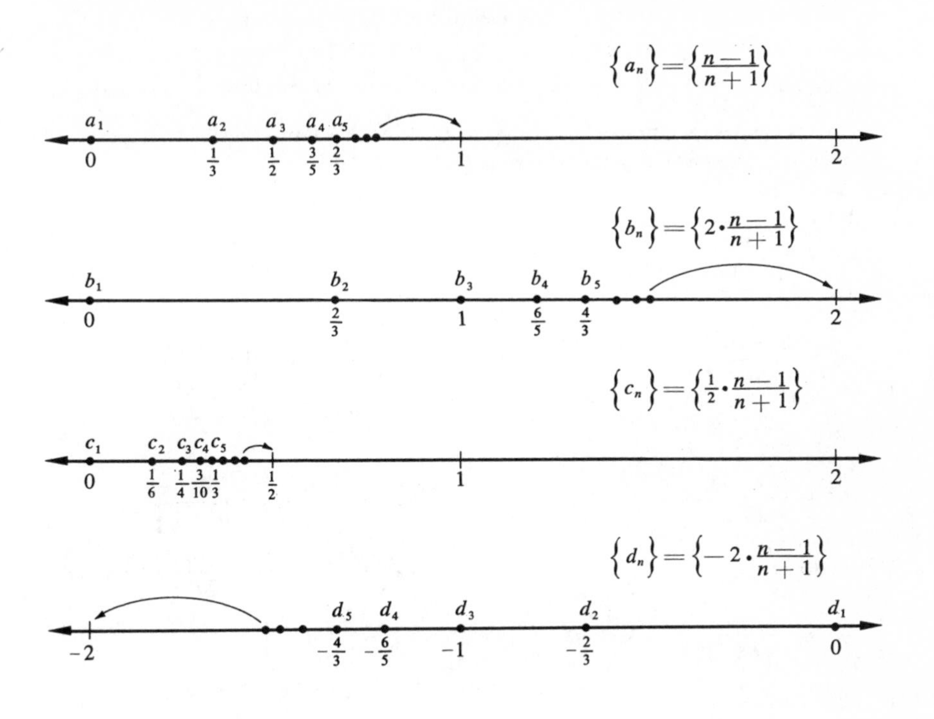

Figure 40

(3) It is easily shown by mathematical induction that Theorems 5-1 and 5-3 can be extended to cover the limit of the sum and the limit of the product of more than two sequences. For example,

$$\{a_n\} = \left\{\frac{6n^2 - 3n + 8}{2n^2}\right\} = \{3\} - \left\{\frac{3}{2}\right\}\left\{\frac{1}{n}\right\} + \{4\}\left\{\frac{1}{n^2}\right\} \to 3 - \tfrac{3}{2}\cdot 0 + 4\cdot 0 = 3$$

$$\{b_n\} = \left\{\frac{(2n^2 + 1)(n + 2)(n + 1)}{n^4}\right\} = \left\{2 + \frac{1}{n^2}\right\}\left\{1 + 2\cdot\frac{1}{n}\right\}\left\{1 + \frac{1}{n}\right\}$$
$$\to (2 + 0)(1 + 2\cdot 0)(1 + 0) = 2$$

(4) Theorem 5-3 also applies to *powers* of sequences. If, for example, $\{a_n\} = \left\{\frac{2n + 1}{n}\right\}$ is multiplied by itself a certain number of times, say five times, then the limit of the resulting sequence is 2^5. Similarly, $\{b_n\} = \left\{\left(\frac{3n}{n + 1}\right)^4\right\} \to 3^4 = 81$.

EXERCISES

1. Assume $\left\{\frac{n}{n+1}\right\} \to 1$, $\{c\} \to c$, and all sequences of the form $\left\{\frac{1}{n^p}\right\}$ (where p is any positive real number) converge to 0. Find the limit, whenever it exists, of each sequence whose general term is given.

$$a_n = \frac{6n+13}{13n} \qquad p_n = \frac{8n}{n+1}$$

$$b_n = \frac{2+3n}{6n} \qquad q_n = \frac{2n+7}{5n}$$

$$c_n = \frac{8-n}{2n} \qquad r_n = \left(\frac{2n+1}{n}\right)\left(5+\frac{7}{n}\right)$$

$$d_n = \frac{n}{n+1} + \frac{n}{2} \qquad s_n = \frac{(2n-1)(3n+4)}{n^2}$$

$$f_n = \frac{5n^2+4}{2n^2} \qquad t_n = \left[\frac{3n}{n+1}\right]^3\left[\frac{2n}{n+1}\right]$$

$$g_n = \frac{2n^2+3n-5}{n^2} \qquad v_n = \frac{(n+1)(n+2)(n+3)(n+4)}{2n^4}$$

$$h_n = \frac{6n-20}{n^2} \qquad w_n = \left[3+\frac{2}{n}\right]^4$$

$$k_n = \frac{n^3}{n^2+20} \qquad y_n = \frac{2\sin^2 n + 2\cos^2 n}{3}$$

2. Let $\{a_n\} = \left\{\frac{n}{n+1}\right\}$ and $\{b_n\} = \left\{\frac{6n}{n+1}\right\}$.

(a) The neighborhood of 6 with radius .1 contains all terms (and only those terms) of $\{b_n\}$ with $n > 59$. What is the radius of the neighborhood of 1 which contains all terms (and only those terms) of $\{a_n\}$ with $n > 59$?

(b) The neighborhood of 6 with radius $\frac{1}{4}$ contains all terms (and only those terms) of $\{b_n\}$ with $n > 23$. What is the radius of the neighborhood of 1 which contains all terms (and only those terms) of $\{a_n\}$ with $n > 23$?

3. Suppose $\{a_n\}$ and $\{b_n\}$ are two sequences such that $\{a_n\} \to A$ and $\{b_n\} = \{5a_n\} \to 5A$, and suppose E is some positive real number such that $\langle A - E, A + E\rangle$ contains exactly those terms of $\{a_n\}$ with $n > 40$. What neighborhood of $5A$ contains exactly those terms of $\{b_n\}$ with $n > 40$?

4. If $\{a_n\}$ and $\{b_n\}$ are two divergent sequences,

(a) is $\{a_n + b_n\}$ always, sometimes, or never convergent? (Give examples.)

(b) is $\{a_nb_n\}$ always, sometimes, or never convergent? (Give examples.)

5. If $\{a_n\}$ is convergent and $\{b_n\}$ is divergent,

(a) is $\{a_n + b_n\}$ always, sometimes, or never convergent?

(b) is $\{a_nb_n\}$ always, sometimes, or never convergent?

(Justify your answers. Give examples.)

6. If two sequences $\{a_n\}$ and $\{b_n\}$ both converge to the same number, what, if anything, can you say about convergence of the sequence $\{a_n - b_n\}$?

Suppose, on the other hand, we have two sequences $\{a_n\}$ and $\{b_n\}$ and we know that $\{a_n - b_n\}$ is a sequence which converges to 0. Does this mean that $\{a_n\}$ and $\{b_n\}$ both converge to the same number?

7. Prove the following, using techniques studied in previous chapters:
(a) any constant sequence $\{s_n\} = \{c\}$ converges to c.
(b) $\{b_n\} = \left\{\frac{1}{n^p}\right\} \to 0$, where p is any positive real number.

2. Proofs of Theorems

Two different proofs of Theorem 5-1 are presented. The first proof is for the special case in which A and B are both positive and $A < B$. The second proof deals with the general case, in which A and B are any two real numbers. The second proof is more formal than the first one but involves essentially the same ideas.

First Proof of Theorem 5-1

This proof is for the special case in which A and B are both positive, all terms of $\{a_n\}$ and $\{b_n\}$ are positive, and $A < B$. Let $a_n + b_n$ be designated by c_n and let $A + B$ be designated by C. Assuming that $\{a_n\} \to A$ and $\{b_n\} \to B$, we wish to prove that $\{c_n\} \to C$. We shall prove this by the definition of convergence, using the general neighborhood $\langle C - E, C + E\rangle$. Choose any neighborhood $\langle C - E, C + E\rangle$ of C. Then choose the neighborhoods $\left\langle A - \frac{E}{2}, A + \frac{E}{2}\right\rangle$ and $\left\langle B - \frac{E}{2}, B + \frac{E}{2}\right\rangle$ of A and B. If x is any number in $\left\langle A - \frac{E}{2}, A + \frac{E}{2}\right\rangle$ and y is any number in $\left\langle B - \frac{E}{2}, B + \frac{E}{2}\right\rangle$, we will show that $x + y$ is in $\langle C - E, C + E\rangle$. Since $x > A - \frac{E}{2}$ and $y > B - \frac{E}{2}$, then $x + y > \left(A - \frac{E}{2}\right) + \left(B - \frac{E}{2}\right)$ and $\left(A - \frac{E}{2}\right) + \left(B - \frac{E}{2}\right) = A + B - E = C - E$. That is, $x + y > C - E$. Similarly, $x < A + \frac{E}{2}$ and $y < B + \frac{E}{2}$, so that $x + y < C + E$. Therefore $x + y$ is in the neighborhood $\langle C - E, C + E\rangle$. (See Figure 41.)

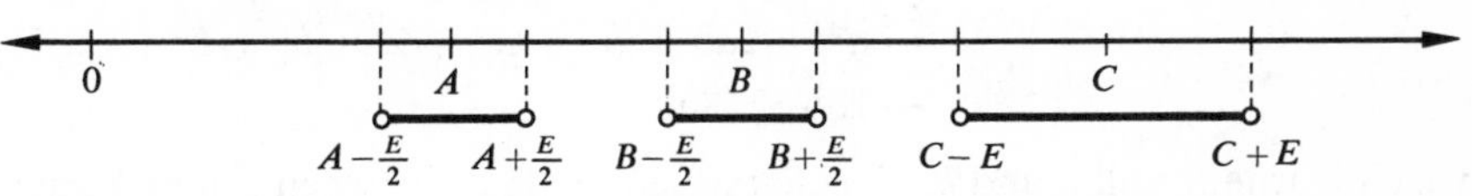

Figure 41

Since $\{a_n\} \to A$, every neighborhood of A contains all but a finite number of terms of $\{a_n\}$, at most. Thus outside the neighborhood $\left\langle A - \frac{E}{2}, A + \frac{E}{2} \right\rangle$ there can be at most a finite number of terms, $a_1, a_2, a_3, \cdots, a_P$, where P is some natural number. (There might be no terms outside this neighborhood.)

Since $\{b_n\} \to B$, outside the neighborhood $\left\langle B - \frac{E}{2}, B + \frac{E}{2} \right\rangle$ there can be at most a finite number of terms, $b_1, b_2, b_3, \cdots, b_Q$, where Q is some natural number. (There might be no terms outside this neighborhood.)

Let M be the larger of the two numbers $P + 1$ and $Q + 1$. Then all terms a_n with $n \geq M$ are in the neighborhood $\left\langle A - \frac{E}{2}, A + \frac{E}{2} \right\rangle$, and all terms b_n with $n \geq M$ are in the neighborhood $\left\langle B - \frac{E}{2}, B + \frac{E}{2} \right\rangle$. Furthermore, from our above results, all terms $c_n = a_n + b_n$ with $n \geq M$ are in the neighborhood $\langle C - E, C + E \rangle$, and this means that $\{c_n\} \to C$. Thus the theorem has been proved.

Before proving this theorem for the general case, three remarks are essential.

REMARK 1: The geometrical statement "x is in the neighborhood $\langle A - E, A + E \rangle$ of A" means, analytically, that the number $x - A$ is less than E in absolute value. This analytical statement is written as "$|x - A| < E$."

For instance, if x is any point in the neighborhood $\langle 2, 4 \rangle$, then $|x - 3| < 1$. In particular, $|3.9 - 3| = |.9| = .9$, $|3.1 - 3| = |.1| = .1$, $|2.9 - 3| = |-.1| = .1$, $|2.5 - 3| = |-.5| = .5$, etc.

As another example, if x is a real number such that $|x - 2| < \frac{1}{2}$, then x lies in the neighborhood $\langle 2 - \frac{1}{2}, 2 + \frac{1}{2} \rangle = \langle 1.5, 2.5 \rangle$ — that is, the neighborhood of 2 of radius $\frac{1}{2}$. Similarly, $|x - (-1)| < \frac{1}{2}$ means that x lies in the neighborhood of (-1) of radius $\frac{1}{2}$ — that is, $|x + 1| < \frac{1}{2}$ means that x lies in the neighborhood $\langle -1\frac{1}{2}, -\frac{1}{2} \rangle$.

REMARK 2: If a and b are any two real numbers, then the inequality $|a + b| \leq |a| + |b|$ is true. That is, for certain pairs of real numbers a and b, $|a + b| = |a| + |b|$, and for other pairs, $|a + b| < |a| + |b|$. In any event, $|a| + |b|$ is never less than $|a + b|$.

We may prove $|a + b| \leq |a| + |b|$ as follows, recalling that for x and y positive, $x \leq y \leftrightarrow x^2 \leq y^2$.

$$
\begin{aligned}
|a + b| \leq |a| + |b| &\leftrightarrow |a + b|^2 \leq (|a| + |b|)^2 \\
&\leftrightarrow (a + b)^2 \leq (|a| + |b|)^2 \\
&\leftrightarrow a^2 + 2ab + b^2 \leq a^2 + 2|a|\,|b| + b^2 \\
&\leftrightarrow 2ab \leq 2|a|\,|b|,
\end{aligned}
$$

which is true for all a and b, since either $ab = -|a|\,|b|$ (when a and b have opposite signs) or $ab = |a|\,|b|$ (otherwise).

REMARK 3: If a and b are any two real numbers, then $|a \cdot b| = |a| \cdot |b|$. For example, $|-2x| = |-2| \cdot |x| = 2 \cdot |x|$, for all real numbers x.

Second Proof of Theorem 5-1

Let $a_n + b_n = c_n$ and $A + B = C$. Choose $\langle C - E, C + E \rangle$ to be *any* neighborhood of C. We wish to show that there is a natural number M such that, for all natural numbers $n \geq M$, c_n is in this neighborhood.

Since $\{a_n\} \to A$, for the neighborhood $\left\langle A - \frac{E}{2}, A + \frac{E}{2} \right\rangle$ there is a natural number M_1 such that, for all $n \geq M_1$, a_n is in this neighborhood. Since $\{b_n\} \to B$, for the neighborhood $\left\langle B - \frac{E}{2}, B + \frac{E}{2} \right\rangle$ there is a natural number M_2 such that, for all $n \geq M_2$, b_n is in this neighborhood.

Let $M =$ the larger of M_1 and M_2. Then for all natural numbers $n \geq M$, a_n is in $\left\langle A - \frac{E}{2}, A + \frac{E}{2} \right\rangle$ and b_n is in $\left\langle B - \frac{E}{2}, B + \frac{E}{2} \right\rangle$.

In analytical language this means that, for all natural numbers $n \geq M$, $|a_n - A| < \frac{E}{2}$ and $|b_n - B| < \frac{E}{2}$.

By *Remark 2* preceding this proof,

$$|(a_n - A) + (b_n - B)| \leq |a_n - A| + |b_n - B|.$$

That is,

$$|(a_n + b_n) - (A + B)| \leq |a_n - A| + |b_n - B|,$$

or

$$|c_n - C| \leq |a_n - A| + |b_n - B|.$$

Furthermore, since

$$|a_n - A| < \frac{E}{2} \text{ and } |b_n - B| < \frac{E}{2},$$

then

$$|a_n - A| + |b_n - B| < \frac{E}{2} + \frac{E}{2} = E.$$

Therefore, $|c_n - C| < E$ for all $n \geq M$. This means, geometrically, that all terms c_n with $n \geq M$ are in the neighborhood $\langle C - E, C + E \rangle$. Thus $\{c_n\} \to C$; that is, $\{a_n + b_n\} \to A + B$.

A proof of Theorem 5-2 would be quite similar to a proof of Theorem 5-1 and is therefore not included. The theorem should be proved as an Exercise, and the following material might be helpful in the proof.

A proof of Theorem 5-2 would be likely to involve the inequality $|(a_n - b_n) - (A - B)| < E$. To obtain this, we would write

$$|(a_n - A) + (B - b_n)| \leq |a_n - A| + |B - b_n|$$
$$= |a_n - A| + |b_n - B| < \frac{E}{2} + \frac{E}{2} = E.$$

But,

$$|(a_n - A) + (B - b_n)| = |(a_n - b_n) - (A - B)|.$$

Therefore,

$$|(a_n - b_n) - (A - B)| < E,$$

and if $d_n = a_n - b_n$ and $D = A - B$, then we have shown that $|d_n - D| < E$ for all $n \geq M$.

A Proof of Theorem 5-3

(For convenience, we are proving the special case when A and B and all terms of $\{a_n\}$ and of $\{b_n\}$ are positive. The general case — that is, without these restrictions — would be very similar and would require only the inclusion of certain absolute value symbols.)

Assume $\{a_n\} \to A$ and $\{b_n\} \to B$. Let $\langle AB - E, AB + E\rangle$ be any neighborhood of AB. We wish to show that there is a natural number M such that, for all natural numbers $n \geq M$, a_nb_n is in this neighborhood.

Since $\{b_n\}$ converges, it is bounded above, so we can find a positive real number P which is $\geq$ all terms of $\{b_n\}$.

For the neighborhood $\left\langle A - \frac{E}{P + A}, A + \frac{E}{P + A}\right\rangle$, there is a natural number M_1 such that, for all $n \geq M_1$, a_n is in this neighborhood.

For the neighborhood $\left\langle B - \frac{E}{P + A}, B + \frac{E}{P + A}\right\rangle$, there is a natural number M_2 such that, for all $n \geq M_2$, b_n is in this neighborhood.

Let M be the greater of the two natural numbers M_1 and M_2. Then for all $n \geq M$,

$$a_n \text{ is in } \left\langle A - \frac{E}{P + A}, A + \frac{E}{P + A}\right\rangle,$$

and

$$b_n \text{ is in } \left\langle B - \frac{E}{P + A}, B + \frac{E}{P + A}\right\rangle.$$

Using absolute value notation, this means that for all $n \geq M$,

$$|a_n - A| < \frac{E}{P + A} \quad \text{and} \quad |b_n - B| < \frac{E}{P + A}.$$

Our task now is to show that $|a_nb_n - AB| < E$ for all $n \geq M$.

$$\begin{aligned} |a_nb_n - AB| &= |a_nb_n - Ab_n + Ab_n - AB| \\ &= |b_n(a_n - A) + A(b_n - B)| \\ &\leq |b_n(a_n - A)| + |A(b_n - B)| \\ &= |b_n|\,|a_n - A| + |A|\,|b_n - B|. \end{aligned}$$

Since $b_n \leq P$, $|a_n - A| < \dfrac{E}{P + A}$, and $|b_n - B| < \dfrac{E}{P + A}$, then the last expression above is less than $P\left(\dfrac{E}{P + A}\right) + A\left(\dfrac{E}{P + A}\right) = (P + A)\left(\dfrac{E}{P + A}\right) = E.$

That is, $|a_nb_n - AB| < E.$ This concludes the proof.

3. Quotients of Sequences

DEFINITION

If $\{a_n\}$ and $\{b_n\}$ are any two sequences and no term of $\{b_n\}$ is 0, then the *quotient* of $\{a_n\}$ and $\{b_n\}$ is defined to be the sequence whose nth term is $\dfrac{a_n}{b_n}$. In notation, $\dfrac{\{a_n\}}{\{b_n\}} = \left\{\dfrac{a_n}{b_n}\right\}$.

Now, given a sequence, if we can express this sequence as the quotient of two convergent sequences, then the quotient of their limits will be the limit of the given sequence. This idea is presented in the following theorem. A proof of the theorem is not included, as it would be quite difficult at this point.

THEOREM 5-4

If $\{a_n\} \to A$ and $\{b_n\} \to B$ $(B \neq 0)$ and if no term of $\{b_n\}$ is 0, then the quotient sequence $\left\{\dfrac{a_n}{b_n}\right\}$ converges to $\dfrac{A}{B}$.

For example, to find the limit of $\{s_n\} = \left\{\dfrac{2n^2 + 3n - 5}{3n^2 - n + 4}\right\}$, we can multiply the general term by $\dfrac{n^{-2}}{n^{-2}}$, obtaining $\left\{\dfrac{2 + \frac{3}{n} - \frac{5}{n^2}}{3 - \frac{1}{n} + \frac{4}{n^2}}\right\}$, and hence $\dfrac{\left\{2 + \frac{3}{n} - \frac{5}{n^2}\right\}}{\left\{3 - \frac{1}{n} + \frac{4}{n^2}\right\}}$.

Then

$$\left\{2+\frac{3}{n}-\frac{5}{n^2}\right\} = \{2\}+\{3\}\left\{\frac{1}{n}\right\}-\{5\}\left\{\frac{1}{n^2}\right\}\to 2+3\cdot 0-5\cdot 0 = 2,$$

and

$$\left\{3-\frac{1}{n}+\frac{4}{n^2}\right\} = \{3\}-\left\{\frac{1}{n}\right\}+\{4\}\left\{\frac{1}{n^2}\right\}\to 3-0+4\cdot 0 = 3.$$

Thus $\{s_n\}\to\frac{2}{3}$.

Although the theorems on sums, differences, and products (Theorems 5-1, 5-2, and 5-3) provide valuable techniques for finding limits, we would be severely handicapped without Theorem 5-4, as the preceding example illustrates. Normally, various combinations of these theorems are applied.

Now that we have these four theorems as tools, we can rapidly determine the limit of a great many types of sequences. For this reason, we have reached a high point in our study of sequences and limits.

Applications of Theorems 5-1, 5-2, 5-3, and 5-4

(1) $$\{a_n\} = \left\{\frac{2n+1}{3n+4}\right\} = \left\{\frac{2+\frac{1}{n}}{3+\frac{4}{n}}\right\} = \frac{\left\{2+\frac{1}{n}\right\}}{\left\{3+\frac{4}{n}\right\}}.$$

By Theorem 5-1,

$$\left\{2+\frac{1}{n}\right\} = \{2\}+\left\{\frac{1}{n}\right\}\to 2+0 = 2 \text{ and}$$

$$\left\{3+\frac{4}{n}\right\} = \{3\}+\left\{\frac{4}{n}\right\}\to 3+0 = 3.$$

Thus by Theorem 5-4, $\{a_n\} = \left\{\dfrac{2n+1}{3n+4}\right\}\to\dfrac{2}{3}$.

(2) Formerly we assumed $\{b_n\} = \left\{\dfrac{n}{n+1}\right\}\to 1$. (See page 107.) Now we are able to prove this without recourse to the definition of convergence.

$$\left\{\frac{n}{n+1}\right\} = \left\{\frac{1}{1+\frac{1}{n}}\right\} = \frac{\{1\}}{\{1\}+\left\{\frac{1}{n}\right\}}, \text{ so that } \left\{\frac{n}{n+1}\right\}\to\frac{1}{1+0} = 1.$$

Similarly, $\{c_n\} = \left\{\dfrac{n}{n+3}\right\} = \dfrac{\{1\}}{\left\{1+\frac{3}{n}\right\}}\to\dfrac{1}{1} = 1.$

$$(3)\ \{d_n\} = \left\{\frac{3n^2 + n - 1}{n^2 + 1}\right\} = \left\{\frac{3 + \frac{1}{n} - \frac{1}{n^2}}{1 + \frac{1}{n^2}}\right\} = \frac{\left\{3 + \frac{1}{n} - \frac{1}{n^2}\right\}}{\left\{1 + \frac{1}{n^2}\right\}}$$

$$= \frac{\{3\} + \left\{\frac{1}{n}\right\} - \left\{\frac{1}{n^2}\right\}}{\{1\} + \left\{\frac{1}{n^2}\right\}}, \text{ so } \{d_n\} \to \frac{3 + 0 - 0}{1 + 0} = 3.$$

EXERCISES

1. Find the limit, whenever it exists, of each sequence whose general term is given. No proof is required.

$$a_n = \frac{3n}{n+3}$$

$$b_n = \frac{n^2}{n^3+4}$$

$$c_n = \frac{n^3}{n^2+40}$$

$$d_n = \frac{13n^2 + 9n - 5}{2n^2 - 7n + 3}$$

$$e_n = n + \frac{1}{n}$$

$$f_n = \frac{n^2 + n}{n^2 - 2}$$

$$g_n = \frac{2(n-4)^3}{n^3}$$

$$h_n = \frac{(n-1)(2n-1)(3n-1)}{(n+1)(3n+4)(5n+7)}$$

$$j_n = \frac{(n+1)(n+2)(n+3)(n+4)}{n^4}$$

$$k_n = \frac{\sqrt{n}+1}{n+1}$$

$$p_n = \frac{(5n+4)^3}{n^3}\left(\frac{8n+7}{5}\right)$$

$$q_n = \frac{n}{\sqrt{n}+1}$$

$$r_n = \frac{1}{n^2}\left[\frac{n(n+1)}{2}\right]$$

$$s_n = \frac{5}{n^3}\left[\frac{n(n+1)(2n+1)}{6}\right]$$

$$t_n = \frac{2}{n^4}\left[\frac{n(n+1)}{2}\right]^2$$

$$u_n = 3\left(\frac{n+1}{n}\right)\left(\frac{2n+1}{n}\right)$$

$$v_n = \left(2 + \frac{1}{n}\right)\left(3 + \frac{1}{n}\right)\left(4 + \frac{1}{n}\right)$$

$$w_n = \left(\frac{2n+5}{n}\right)^2\left(5 - \frac{1}{n}\right)$$

2. Let a_n be the largest nth-place decimal whose square is less than 3. (The first four terms are 1.7, 1.73, 1.732, and 1.7320.)

Let b_n be the largest nth-place decimal whose square is less than 6. (The first four terms are 2.4, 2.44, 2.449, and 2.4494.)

(a) If $\{c_n\} = \{a_n + b_n\}$, find c_1, c_2, c_3, and c_4. What number is the limit of $\{c_n\}$, or does $\{c_n\}$ diverge?

(b) If $\{d_n\} = \{a_n b_n\}$, find d_1, d_2, d_3, and d_4. What number is the limit of $\{d_n\}$, or does $\{d_n\}$ diverge?

(c) Find the first four terms of $\{h_n\} = \left\{\frac{b_n}{a_n}\right\}$. What number seems to be the limit of this sequence?

(d) A certain student has a rather unusual mathematical background. He has a thorough understanding of the notions of sequences and convergence, but no knowledge at all of irrational numbers. How might you explain to him why $\sqrt{3} + \sqrt{5}$ is *not* equal to $\sqrt{8}$, whereas $\sqrt{3} \cdot \sqrt{5}$ *is* equal to $\sqrt{15}$? Also, what can be said about the value of $\frac{\sqrt{3}}{\sqrt{5}}$?

4. Domination Principle

The Domination Principle provides another method for proving convergence.

THEOREM 5-5

Domination Principle. If two sequences $\{a_n\}$ and $\{b_n\}$ both converge to the same number L and if $\{c_n\}$ is a sequence such that $a_n \leq c_n \leq b_n$ for all n greater than or equal to some natural number M, then $\{c_n\}$ also converges to L.

The following examples illustrate the use of the Domination Principle in proving convergence.

Examples **(1)** Since $\{a_n\} = \{0\} \to 0$ and $\{b_n\} = \left\{\frac{1}{n}\right\} \to 0$, and since

$$0 \leq \frac{1}{n+1} \leq \frac{1}{n} \leftrightarrow \frac{1}{n+1} \leq \frac{1}{n}$$

$\leftrightarrow n + 1 \geq n$, which is true for all natural numbers,

then by the Domination Principle $\{c_n\} = \left\{\frac{1}{n+1}\right\} \to 0$.

(2) Assume that $\{a_n\} = \left\{\frac{n}{n+1}\right\} \to 1$ and $\{b_n\} = \left\{\frac{n+1}{n}\right\} \to 1$.

Then $\frac{n}{n+1} \leq \frac{n^2+10}{n^2+n} \leq \frac{n+1}{n} \leftrightarrow \frac{n}{n+1} \leq \frac{n^2+10}{n^2+n}$ and $\frac{n^2+10}{n^2+n} \leq \frac{n+1}{n}$:

$$\frac{n}{n+1} \leq \frac{n^2+10}{n^2+n} \leftrightarrow n^3+n^2 \leq n^3+n^2+10n+10$$
$$\leftrightarrow 0 \leq 10n+10, \text{ which is true for all natural numbers } n;$$

and

$$\frac{n^2+10}{n^2+n} \leq \frac{n+1}{n} \leftrightarrow n^3+10n \leq n^3+2n^2+n$$
$$\leftrightarrow 9n \leq 2n^2$$
$$\leftrightarrow \tfrac{9}{2} \leq n, \text{ which is true for all natural numbers } \geq 5.$$

Thus if $\{c_n\} = \left\{\frac{n^2+10}{n^2+n}\right\}$, then for all natural numbers ≥ 5, $a_n \leq c_n \leq b_n$. So $\{c_n\} \to 1$ by the Domination Principle.

EXERCISES

1. Find the limit, whenever it exists. Use any method you wish, but try first to use the Domination Principle.

$$a_n = \frac{1}{2^n} \qquad g_n = \frac{\sin \frac{n}{4}\pi}{n}$$

$$b_n = \frac{50}{n^2} \qquad h_n = \frac{n + \sin \frac{n}{2}\pi}{n + \cos \frac{n}{2}\pi}$$

$$c_n = \frac{\sin n}{n}$$

$$d_n = \frac{1 + \sin n}{n} \qquad k_n = \sqrt{\frac{4n}{n+1}}$$

$$f_n = \frac{\sin n + \cos n}{2n} \qquad p_n = \frac{8 - 6n + 9n^2}{4 - 4n + 3n^2}$$

2. Prove convergence or divergence of each of the sequences whose general term is given. Use any method.

$$a_n = \frac{2^n}{n^n} \qquad b_n = \frac{n^n}{2^n} \qquad c_n = \frac{n!}{2^n} \qquad d_n = \frac{2^n}{n!}$$

6 • Applications: Sequences of Partial Sums

1. Sequences of Partial Sums

If $\{a_n\}$ is a sequence, then a new sequence, which we shall call $\{S_n\}$, can be created by letting

$$S_1 = a_1,$$

$$S_2 = a_1 + a_2,$$

$$S_3 = a_1 + a_2 + a_3,$$

and, in general, $$S_n = a_1 + a_2 + a_3 + \cdots + a_n.$$

This new sequence can be described by $\{S_n\} = \{a_1 + a_2 + a_3 + \cdots + a_n\}$. We shall refer to it as *the sequence of partial sums corresponding to* $\{a_n\}$.

Rather than use small letters of the alphabet, as in b_n, c_n, s_n, t_n, to name sequences of partial sums, it will be helpful to use capital letters instead, such as B_n, C_n, S_n, T_n.

Examples **(1)** The first four terms of $\{S_n\} = \{1 + \frac{1}{2} + \frac{1}{4} + \frac{1}{8} + \cdots + (\frac{1}{2})^{n-1}\}$ are 1, $1 + \frac{1}{2}$, $1 + \frac{1}{2} + \frac{1}{4}$, $1 + \frac{1}{2} + \frac{1}{4} + \frac{1}{8}$, which, when simplified, are 1, $1\frac{1}{2}$, $1\frac{3}{4}$, and $1\frac{7}{8}$.

(2) The first five terms of $\{T_n\} = \{1 - \frac{1}{3} + \frac{1}{9} - \frac{1}{27} + \cdots + (-\frac{1}{3})^{n-1}\}$ are $1, \frac{2}{3}, \frac{7}{9}, \frac{20}{27}, \frac{61}{81}$; or $\frac{81}{81}, \frac{54}{81}, \frac{63}{81}, \frac{60}{81}, \frac{61}{81}$.

(3) Let $\{a_n\} = \left\{\dfrac{2n-1}{(n^2+1)(n^2-2n+2)}\right\}$. Then the first five terms of

$$\{V_n\} = \{a_1 + a_2 + a_3 + \cdots + a_n\} =$$

$$\left\{\tfrac{1}{2} + \tfrac{3}{10} + \tfrac{1}{10} + \tfrac{7}{170} + \tfrac{9}{442} + \cdots + \frac{2n-1}{(n^2+1)(n^2-2n+2)}\right\}$$

are $\frac{1}{2}$, $\frac{4}{5}$, $\frac{9}{10}$, $\frac{16}{17}$ and $\frac{25}{26}$.

(4) The first ten terms of $\{W_n\} = \{32 - 16 + 8 - 4 + \cdots + 32(-\frac{1}{2})^{n-1}\}$ are 32, 16, 24, 20, 22, 21, $21\frac{1}{2}$, $21\frac{1}{4}$, $21\frac{3}{8}$ and $21\frac{5}{16}$.

General Term For a Sequence of Partial Sums

It is sometimes helpful to find a single general term representing the nth partial sum of a given sequence of partial sums.

Any geometric sequence of partial sums $\{a + ar + ar^2 + \cdots + ar^{n-1}\}$, $r \neq 1$, can be described using the general term $a\left(\frac{1-r^n}{1-r}\right)$. Thus

$$\{S_n\} = \{a + ar + ar^2 + \cdots + ar^{n-1}\} = \left\{a\left(\frac{1-r^n}{1-r}\right)\right\}.$$

For example, if $a = 32$ and $r = -\frac{1}{2}$, then

$$\{S_n\} = \{32 - 16 + 8 - 4 + \cdots + 32(-\tfrac{1}{2})^{n-1}\} = \left\{32\left[\frac{1-(-\frac{1}{2})^n}{1-(-\frac{1}{2})}\right]\right\}$$
$$= \{32 \cdot \tfrac{2}{3}[1 - (-\tfrac{1}{2})^n]\}.$$

Also, if $a = 10$ and $r = 2$, then

$$\{R_n\} = \{10 + 20 + 40 + 80 + \cdots + 10(2)^{n-1}\} = \left\{10\left[\frac{1-2^n}{1-2}\right]\right\}$$
$$= \{10(2^n - 1)\}.$$

To find the sum $10 + 20 + 40 + 80 + 160$, we find the fifth term of $\{R_n\}$, which is $10(2^5 - 1) = 10(31) = 310$.

To prove that $S_n = a + ar + ar^2 + \cdots + ar^{n-1}$ is equal to $a\left(\frac{1-r^n}{1-r}\right)$ for $r \neq 1$, we multiply both S_n and $a + ar + ar^2 + \cdots + ar^{n-1}$ by r, obtaining

$$rS_n = ar + ar^2 + ar^3 + \cdots + ar^{n-1} + ar^n.$$

Since $S_n = a + ar + ar^2 + \cdots + ar^{n-1}$, then

$$rS_n - S_n = -a + ar^n \leftrightarrow S_n(r-1) = a(r^n - 1)$$
$$\leftrightarrow S_n = a\left(\frac{r^n-1}{r-1}\right) = a\left(\frac{1-r^n}{1-r}\right),\ r \neq 1.$$

For the sequence of partial sums $\{T_n\} = \{1 + 2 + 3 + \cdots + n\}$, the nth term is given by $\dfrac{n+1}{2} \cdot n$. Let us illustrate why this is true, using $n = 7$:

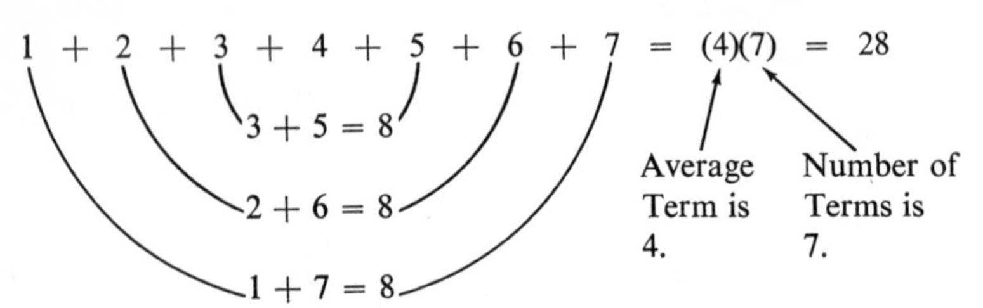

For $n = 7$, the expression $\dfrac{n+1}{2}$ is 4, and 4 is the average term in the expression $1 + 2 + 3 + 4 + 5 + 6 + 7$. Since there are seven terms and each has an average value of 4, then their sum is $4 \cdot 7 = 28$. Thus the sequence of partial sums $\{T_n\} = \{1 + 2 + 3 + \cdots + n\}$ is the same sequence as $\{t_n\} = \left\{\dfrac{n+1}{2} \cdot n\right\}$.

The following three statements are also true:

(1) The sequence $\{A_n\} = \{1 + 3 + 5 + \cdots + (2n - 1)\}$ is the same sequence as $\{a_n\} = \{n^2\}$. (See Figure 42.)

(2) The sequence $\{B_n\} = \{1^2 + 2^2 + 3^2 + \cdots + n^2\}$ is the same sequence as $\{b_n\} = \left\{\dfrac{n(n+1)(2n+1)}{6}\right\}$.

(3) The sequence $\{C_n\} = \{1^3 + 2^3 + 3^3 + \cdots + n^3\}$ is the same sequence as $\{c_n\} = \left\{\left[\dfrac{n(n+1)}{2}\right]^2\right\}$.

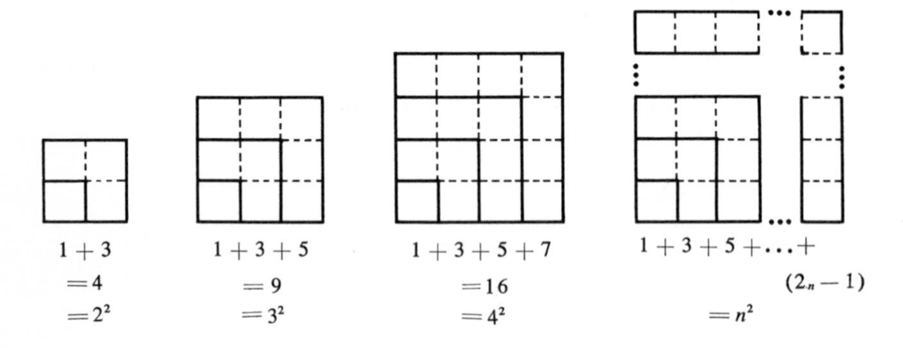

Figure 42

Discoveries of these and other general terms representing sequences of partial sums have involved various kinds of mathematical experimentations over the centuries. Some of the world's foremost mathematicians have been responsible for a number of the discoveries. Once an expression has been discovered which appears to be a general expression for denoting a certain sequence of partial sums, its validity for all natural numbers must be proved. Perhaps the most common method of proof is mathematical induction. Sometimes an informal proof is possible, such as the ones given for

$$\{1 + 2 + 3 + \cdots + n\} = \left\{\frac{n+1}{2} \cdot n\right\}$$

in the preceding discussion, and for

$$\{1 + 3 + 5 + \cdots + (2n - 1)\} = \{n^2\}$$

in Figure 42, and the one required in Problem 4 of the next exercise set.

Convergence of Sequences of Partial Sums

By finding a general term to represent a given sequence of partial sums, we may be able to use this general term to determine convergence or divergence of the sequence of partial sums.

Examples **(1)** Since $\{R_n\} = \{1 + 3 + 5 + \cdots + (2n - 1)\}$ is the same sequence as $\{r_n\} = \{n^2\}$, we can use $\{r_n\}$ to show divergence of $\{R_n\}$. We know by Theorem 3-3(a) that $\{r_n\} = \{n^2\}$ increases without bound and hence diverges.

(2) The sequence of partial sums

$$\{W_n\} = \left\{\tfrac{1}{2} + \tfrac{1}{6} + \tfrac{1}{12} + \tfrac{1}{20} + \cdots + \frac{1}{n(n+1)}\right\}$$

is the same sequence as $\{w_n\} = \left\{\dfrac{n}{n+1}\right\}$, and hence converges to 1. Notice that

$$\tfrac{1}{2} = \tfrac{1}{2} \qquad \tfrac{1}{2} + \tfrac{1}{6} + \tfrac{1}{12} = \tfrac{3}{4}$$

$$\tfrac{1}{2} + \tfrac{1}{6} = \tfrac{2}{3} \qquad \tfrac{1}{2} + \tfrac{1}{6} + \tfrac{1}{12} + \tfrac{1}{20} = \tfrac{4}{5},$$

and $\frac{1}{2}$, $\frac{2}{3}$, $\frac{3}{4}$, and $\frac{4}{5}$ are the first four terms of $\{w_n\} = \left\{\dfrac{n}{n+1}\right\}$.

(3) Any geometric sequence of partial sums

$$\{S_n\} = \{a_1 + a_1r + a_1r^2 + \cdots + a_1r^{n-1}\}$$

equals $\left\{a_1\left(\dfrac{1-r^n}{1-r}\right)\right\}$ for $r \neq 1$, and converges to $a_1\left(\dfrac{1-0}{1-r}\right) = \dfrac{a_1}{1-r}$ *if and only if* $0 < |r| < 1$. This convergence results from the fact that if $0 < |r| < 1$ (that is, $-1 < r < 0$ or $0 < r < 1$), then $\{r^n\} \to 0$, and if $|r| > 1$, $\{r^n\}$ diverges. For $|r| = 1$: if $r = 1$,

$$\{S_n\} = \{a_1 + a_1r + a_1r^2 + \cdots + a_1r^{n-1}\} = \{a_1n\};$$

if $r = -1$, the terms of $\{S_n\}$ are alternately a_1 and 0. In either event, $\{S_n\}$ diverges.

$$\text{Thus, } \{A_n\} = \{1 + \tfrac{1}{2} + \tfrac{1}{4} + \cdots + (\tfrac{1}{2})^{n-1}\} \to \frac{1}{1 - \frac{1}{2}} = 2,$$

$$\{B_n\} = \{1 - \tfrac{1}{3} + \tfrac{1}{9} - \cdots + (-\tfrac{1}{3})^{n-1}\} \to \frac{1}{1 - (-\frac{1}{3})} = \tfrac{3}{4}, \text{ and}$$

$$\{C_n\} = \{25 + 5 + 1 + \cdots + 25(\tfrac{1}{5})^{n-1}\} \to \frac{25}{1 - \frac{1}{5}} = 31\tfrac{1}{4}.$$

If there is no known general term to represent a given sequence of partial sums, there are still numerous methods available for determining convergence or divergence. The main purpose of the next chapter is to study a number of these methods in detail. In the remaining sections of this chapter we will deal with applications of sequences of partial sums to problems involving *areas* of various regions, *volumes* of various solid geometrical figures, and the concept of *work*.

EXERCISES

1. Compute the first three terms of each of the following sequences of partial sums and find its limit, whenever it exists.

$$\{S_n\} = \{49 + .49 + .0049 + \cdots + (49)(.01)^{n-1}\}$$
$$\{T_n\} = \{.7 + .07 + .007 + \cdots + (.7)(.1)^{n-1}\}$$
$$\{V_n\} = \{1 + 2 + 4 + \cdots + 2^{n-1}\}$$
$$\{W_n\} = \{1 - \tfrac{1}{2} + \tfrac{1}{4} - \tfrac{1}{8} + \cdots + (-\tfrac{1}{2})^{n-1}\}$$
$$\{P_n\} = \{2 + \tfrac{2}{3} + \tfrac{2}{9} + \cdots + 2(\tfrac{1}{3})^{n-1}\}$$

2. A basketball was thrown straight up from a point on the floor to a height of 48 feet and on each rebound rose three-fourths of its previous height.
(a) Devise a sequence of partial sums whose limit is the total distance the ball would travel *if* it continued bouncing forever. Find this limit.
(b) If the ball stopped movement upon returning to the ground the fourteenth time, which of the following represents the total distance which the ball traveled?

(i) $48 \cdot 4[1 - (\frac{3}{4})^{13}]$ (iii) $48 \cdot 4[1 - (\frac{3}{4})^{14}]$

(ii) $48 \cdot 8[1 - (\frac{3}{4})^{13}]$ (iv) $48 \cdot 8[1 - (\frac{3}{4})^{14}]$

3. Match each sequence in the left-hand column with the sequence in the right-hand column which is equal to it. (There are more sequences in the right-hand column than are needed.)

(a) $\{P_n\} = \left\{\frac{1}{3} + \frac{1}{15} + \frac{1}{35} + \cdots + \frac{1}{4n^2 - 1}\right\}$

(b) $\{Q_n\} = \left\{\frac{1}{2} + \frac{5}{6} + \frac{11}{12} + \cdots + \left(1 - \frac{1}{n(n+1)}\right)\right\}$

(c) $\{R_n\} = \left\{\frac{1}{2} + \frac{1}{4} + \frac{1}{8} + \cdots + \frac{1}{2^n}\right\}$

$\{a_n\} = \left\{\frac{n^2}{n+1}\right\}$

$\{b_n\} = \left\{1 - \frac{1}{2^n}\right\}$

$\{c_n\} = \left\{\frac{n}{3n-1}\right\}$

$\{d_n\} = \left\{\frac{n}{2n+1}\right\}$

$\{f_n\} = \left\{1 - \frac{1}{2^{n-1}}\right\}$

4. (a) What generalization can you make regarding the sum $2 + 4 + 6 + \cdots + 2n$, on the basis of the geometrical pattern suggested in Figure 43?
(b) How can you relate the generalization made in (a) to the sum $1 + 2 + 3 + \cdots + n$?

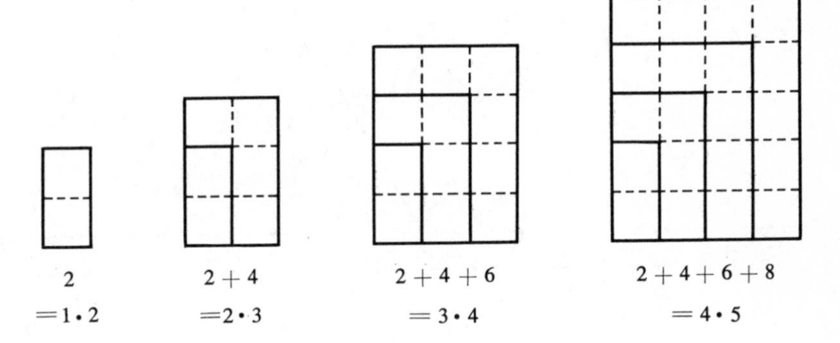

Figure 43

5. If $S_n = 5 + 5(\frac{1}{3}) + 5(\frac{1}{3})^2 + \cdots + 5(\frac{1}{3})^{n-1}$ and the limit of $\{S_n\}$ is one-third the limit of a sequence of partial sums $\{T_n\}$, find one such sequence of partial sums $\{T_n\}$.

6. If $r = 1$ in $\{a_n\} = \{(.7)r^{n-1}\}$, what is the limit, if it exists, of the corresponding sequence of partial sums?

7. In Figure 44, triangle ABC is equilateral, each side being 4 inches long. The midpoints of the sides form the vertices of equilateral triangle DEF. The midpoints of the sides of triangle DEF form the vertices of equilateral triangle KGH. Suppose this process of forming triangles were continued indefinitely. What would be the sum of the perimeters of all the triangles?

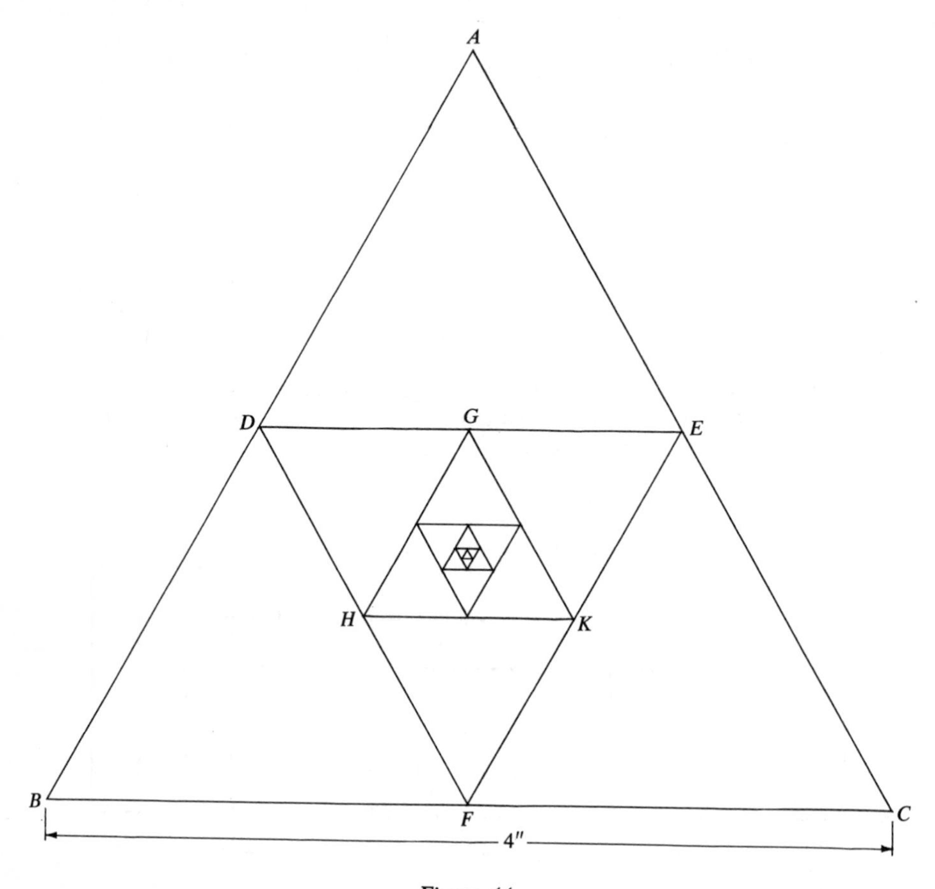

Figure 44

8. *Achilles and the Tortoise.* Zeno, one of the most prominent investigators of problems of sequences of partial sums in the fifth century B.C., argued that Achilles could not pass a tortoise even though he went faster than the tortoise, provided the tortoise was given a head start in the race. He argued that if Achilles could run ten times as fast as the tortoise and if the tortoise had a start of 1000 yards, then when Achilles had gone the 1000 yards, the tortoise would be 100 yards ahead of him. When Achilles had gone those 100 yards, the tortoise would be 10 yards ahead of him. This would continue, and Achilles would come nearer and nearer to the tortoise but would never reach him. Was Zeno right in his reasoning?

2. Area Problems

Many of the early mathematicians, such as Euclid and Archimedes, in response to practical needs such as finding areas bounded by curves, developed solutions to problems in an intuitive fashion. Only within the past century have formal specifications of the theory of limits crystallized. Yet, our modern presentations of the theory of limits have much in common with some of the early presentations. The intuitive attempts of some early mathematicians did reach the heart of the matter.

A number of the problems in this chapter illustrate not only certain kinds of thinking which originated with early mathematicians, but also certain fundamental ideas inherent in contemporary mathematics. Limits play a major role in finding the area of a region bounded by various lines and curves. Your later study of calculus will reveal general techniques for finding areas — techniques based upon the theory of limits. For some initial understanding of these, we will now find areas of certain regions of the coordinate plane.

As a starting point in our discussion of area of planar regions, we make the following assumption:

ASSUMPTION

The area of a rectangle is the product of the lengths of any two consecutive sides.

This assumption is true for *any* rectangle; that is, if a and b, the lengths of two consecutive sides, are any two positive real numbers, the area is ab. In the special event when a and b are natural numbers, the rectangle may be divided into ab unit squares, thus indicating the plausibility of making this assumption.

As is customary in discussions of areas of planar regions, we shall use the above assumption to determine the area of other planar regions. The first such region is the triangular region bounded by the x-axis, the vertical line $x = 1$ and the line $y = x$. (See Figure 45.) Our purpose in determining the area of this triangular region is to illustrate a method which we will use for finding areas of other planar regions.

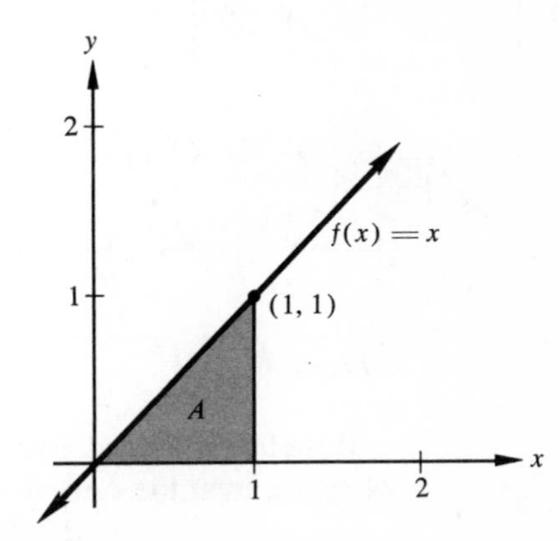

Figure 45

Problem Find the area A of the region bounded by the x-axis, the vertical line $x = 1$ and the graph of $f(x) = x$. (Refer back to Figure 45 on the preceding page.)

Solution: Suppose we divide the interval [0, 1] on the x-axis into eight congruent subintervals, each of width $\frac{1}{8}$. The left end points of these subintervals are $0, \frac{1}{8}, \frac{2}{8}, \frac{3}{8}, \cdots, \frac{7}{8}$. For each of these subintervals (except the first one) we construct a rectangle whose width is $\frac{1}{8}$ and whose height is the f-value of the left end point, where $f(x) = x$.(See Figure 46.)

The heights of the first three rectangles are, respectively, $\frac{1}{8}$, $\frac{2}{8}$, and $\frac{3}{8}$. The sum of the areas of the seven rectangles is

$$\tfrac{1}{8}(\tfrac{1}{8}) + \tfrac{1}{8}(\tfrac{2}{8}) + \tfrac{1}{8}(\tfrac{3}{8}) + \cdots + \tfrac{1}{8}(\tfrac{7}{8}) = \frac{1}{8^2}(1 + 2 + 3 + \cdots + 7).$$

Previously in this chapter it was learned that, for all natural numbers n,

$$1 + 2 + 3 + \cdots + n = \frac{n+1}{2}\cdot n.$$

For $n = 7$ we have $1 + 2 + 3 + \cdots + 7 = 28$. This means that the sum of the areas of the seven rectangles is $\frac{1}{8^2}(28) = \frac{7}{16}\cdot$ This number is an approximation to the area of the shaded region.

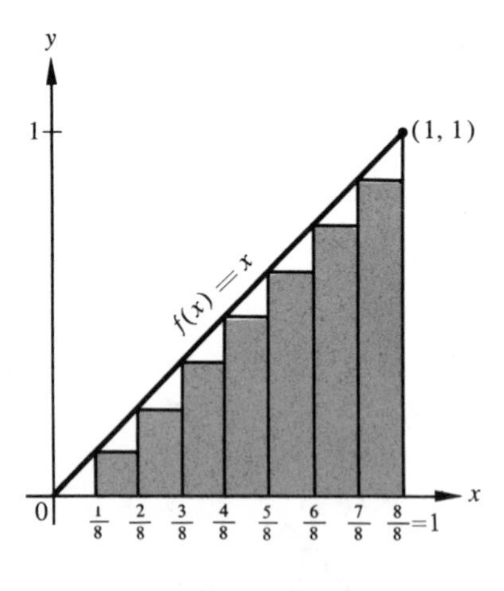

Figure 46

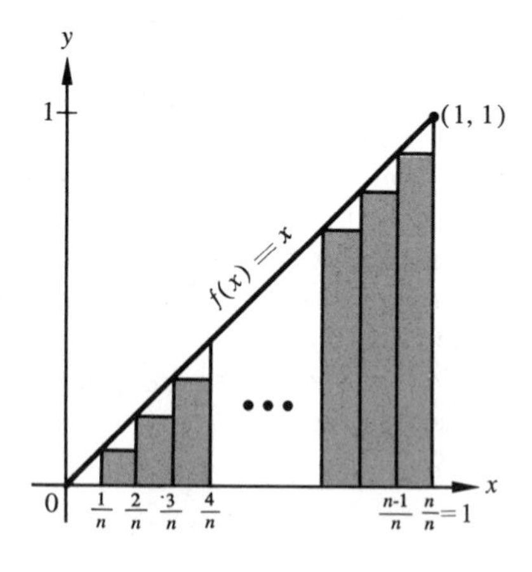

Figure 47

If we had inscribed, say, 999 nonoverlapping rectangles instead of seven, then the sum of all their areas would be a much better approximation to the area of the shaded region. In fact, this number would be

$$\frac{1}{1000}\left(\frac{1}{1000}\right) + \frac{1}{1000}\left(\frac{2}{1000}\right) + \frac{1}{1000}\left(\frac{3}{1000}\right) + \cdots + \frac{1}{1000}\left(\frac{999}{1000}\right) = \frac{1}{10^6}\left[\frac{999+1}{2}\cdot 999\right] = \frac{999}{2000},$$

which differs from $\frac{1}{2}$ by only .0005 and which would be .500 when rounded off to three decimal places.

Let us now divide the interval [0, 1] into n congruent subintervals, where n is any natural number. (See Figure 47.) We shall proceed as we did with $n = 8$ and $n = 1000$. The *left* end points of the n subintervals are $0, \frac{1}{n}, \frac{2}{n}, \frac{3}{n}, \ldots, \frac{n-1}{n}$. For each of these subintervals (except the first one) we construct a rectangle whose width is $\frac{1}{n}$ and whose height is the f-value of the left end point, where $f(x) = x$. For example, the heights of the first three rectangles are $\frac{1}{n}$, $\frac{2}{n}$ and $\frac{3}{n}$, and the height of the last rectangle is $\frac{n-1}{n}$.

The areas of the first three rectangles are

$$\frac{1}{n}\left(\frac{1}{n}\right), \frac{1}{n}\left(\frac{2}{n}\right) \text{ and } \frac{1}{n}\left(\frac{3}{n}\right).$$

We devise a sequence $\{L_n\}$ whose general term, L_n, is the sum of the areas of the $(n-1)$ rectangles. Then

$$\begin{aligned} L_n &= \frac{1}{n}\left(\frac{1}{n}\right) + \frac{1}{n}\left(\frac{2}{n}\right) + \frac{1}{n}\left(\frac{3}{n}\right) + \cdots + \frac{1}{n}\left(\frac{n-1}{n}\right) \\ &= \frac{1}{n^2}[1 + 2 + 3 + \cdots + (n-1)] \\ &= \frac{1}{n^2}\left[\frac{n}{2}(n-1)\right] \text{(using } 1 + 2 + 3 + \cdots + n = \frac{n+1}{2}\cdot n, \\ &\qquad \text{with } n \text{ replaced by } n-1) \\ &= \tfrac{1}{2}\left(1 - \frac{1}{n}\right). \end{aligned}$$

So $\{L_n\} \to \frac{1}{2}$.

Before we discuss the significance of this limit, we shall devise a sequence $\{U_n\}$ as follows.

Divide the interval [0, 1] into n congruent subintervals. The *right* end points of the n subintervals are $\frac{1}{n}, \frac{2}{n}, \frac{3}{n}, \ldots, \frac{n}{n}$. For each of these subintervals, construct a rectangle whose width is $\frac{1}{n}$ and whose height is the f-value of the right end point, where $f(x) = x$. (See Figure 48 on the opposite page.)

Let U_n equal the sum of the areas of the n rectangles. Then

$$U_n = \frac{1}{n}\left(\frac{1}{n}\right) + \frac{1}{n}\left(\frac{2}{n}\right) + \frac{1}{n}\left(\frac{3}{n}\right) + \cdots + \frac{1}{n}\left(\frac{n}{n}\right)$$

$$= \frac{1}{n^2}[1 + 2 + 3 + \cdots + n]$$

$$= \frac{1}{n^2}\left[\frac{n+1}{2} \cdot n\right]$$

$$= \tfrac{1}{2}\left(1 + \frac{1}{n}\right),$$

so $\{U_n\} \to \frac{1}{2}$.

The area A of the region under consideration is greater than every term of $\{L_n\}$, and the terms of $\{L_n\}$ are increasing, with A as an upper bound. Similarly, A is less than every term of $\{U_n\}$, and the terms of $\{U_n\}$ are decreasing, with A as a lower bound. Thus A is $\geq$ the limit of $\{L_n\}$ and A is $\leq$ the limit of $\{U_n\}$. But the limit of both sequences is $\frac{1}{2}$, so that A must be $\frac{1}{2}$.

REMARK: In the preceding example, we devised *two* sequences, $\{L_n\}$ and $\{U_n\}$, both of which converge to $\frac{1}{2}$, which is the area of the given triangular region. We devised *both* sequences in order to point out that they would have the same limit and that hence we could have used either sequence to determine the area. In succeeding problems involving area, we will use only one such sequence, either the sequence of lower sums, $\{L_n\}$, or the sequence of upper sums, $\{U_n\}$.

The method demonstrated in the preceding problem can be used to find areas of other regions of the coordinate plane, as the problem on the opposite page illustrates.

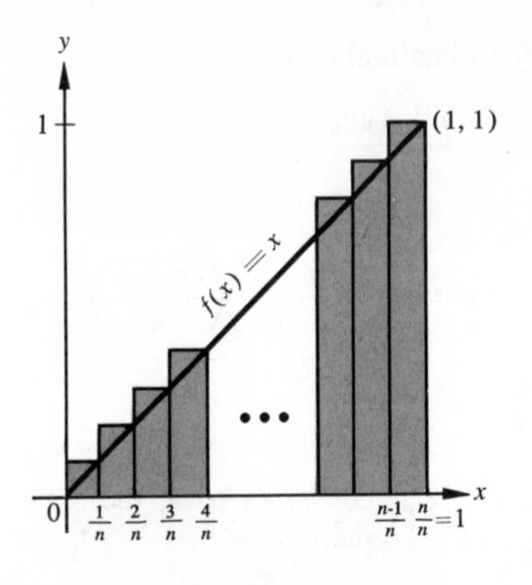

Figure 48

Figure 49

Problem Find the area A of the region of the plane bounded by the x-axis, the vertical line $x = 1$ and the graph of $f(x) = x^2$.

Solution: Divide the interval [0, 1] on the x-axis into n congruent subintervals, where n is any natural number. The right end points of the n subintervals are $\frac{1}{n}, \frac{2}{n}, \frac{3}{n}, \cdots, \frac{n}{n}$. For each of these subintervals we construct a rectangle whose width is $\frac{1}{n}$ and whose height is the f-value of the right end point, where $f(x) = x^2$. For example, the heights of the first three rectangles are

$$f\left(\frac{1}{n}\right) = \left(\frac{1}{n}\right)^2, \quad f\left(\frac{2}{n}\right) = \left(\frac{2}{n}\right)^2 \quad \text{and} \quad f\left(\frac{3}{n}\right) = \left(\frac{3}{n}\right)^2.$$

Then the areas of the first three rectangles are

$$\frac{1}{n}\left(\frac{1}{n}\right)^2, \quad \frac{1}{n}\left(\frac{2}{n}\right)^2, \quad \text{and} \quad \frac{1}{n}\left(\frac{3}{n}\right)^2.$$ (See Figure 49.)

Devise a sequence $\{U_n\}$ whose general term is the sum of the areas of the n rectangles. Then

$$U_n = \frac{1}{n}\left(\frac{1}{n}\right)^2 + \frac{1}{n}\left(\frac{2}{n}\right)^2 + \frac{1}{n}\left(\frac{3}{n}\right)^2 + \cdots + \frac{1}{n}\left(\frac{n}{n}\right)^2$$

$$= \frac{1}{n^3}[1^2 + 2^2 + 3^2 + \cdots + n^2].$$

Previously it was noted that, for all natural numbers n,

$$1^2 + 2^2 + 3^2 + \cdots + n^2 = \frac{n(n+1)(2n+1)}{6},$$

so that

$$U_n = \frac{1}{n^3}\left[\frac{n(n+1)(2n+1)}{6}\right] = \frac{1}{6}\left(\frac{n}{n}\right)\left(\frac{n+1}{n}\right)\left(\frac{2n+1}{n}\right)$$

$$= \frac{1}{6}\left(1+\frac{1}{n}\right)\left(2+\frac{1}{n}\right).$$

Then $\{U_n\} \to \frac{1}{6}\cdot 1\cdot 2 = \frac{1}{3}$. Since $\{U_n\}$ converges to $\frac{1}{3}$, the area of the shaded region in Figure 49 is $\frac{1}{3}$ square units.

EXERCISES

1. Find the area of the region bounded by the curve $y = x^3$, the x-axis, and the vertical line $x = 1$.

2. Find the area of the region bounded by the curve $y = 2x^2$, the x-axis, and the vertical line $x = 1$.

3. Find the area of the region bounded by the curve $y = 2x^3$, the x-axis, and the vertical line $x = 1$.

4. In the discussion in the text and in the three preceding problems, you now should have noticed that the area of the region bounded by the x-axis, the line $x = 1$, and
(a) the curve $y = x^1$ is __?__.
(b) the curve $y = x^2$ is __?__.
(c) the curve $y = x^3$ is __?__.
What pattern seems to be forming? Can you *guess* what the area would be under $y = x^4$? under $y = x^5$? under $y = x^n$, where n is any natural number? What is the area of the region bounded by the x-axis, the line $x = 1$, and
(d) the curve $y = 2x$?
(e) the curve $y = 2x^2$?
(f) the curve $y = 2x^3$?
Can you guess the area under $y = 2x^4$? under $y = 2x^n$, where n is any natural number? under $y = cx^n$, where c is any positive real number and n is any natural number?

5. (a) Find the area of the region bounded by the curve $y = x^2$, the x-axis, and the vertical line $x = 2$.
(b) What is the area of the region bounded by the curve $y = x^2$, the x-axis, and the two vertical lines $x = 1$ and $x = 2$?

3. Volume of a Right Circular Cone

DEFINITION

Let set S consist of a given circle and all the points in the interior of the circle. Let point C be the center of the circle and the number r the radius. Let line L be the straight line which passes through C and which is perpendicular to the plane containing the circle. Let V be any specified point on line L except C. Then the union of all line segments QV, where Q represents any point in set S, is called a *right circular cone* with base S, vertex V, and radius r. The altitude of the cone is the length of line segment VC and is referred to as h. (See Figure 50.)

Our purpose now is to develop a formula for the volume of a right circular cone, in the special case when $h = 1$. The radius r of the circular base can be any positive real number. The cone is assumed to be positioned as follows: The vertex V is placed at the point $(0, 0)$, and point C (the center of the circular base) is placed at the point $(1, 0)$ on the x-axis in the coordinate plane.

First we divide the interval $[0, 1]$ on the x-axis into n congruent subintervals, each of width $\frac{1}{n}$, where n is any natural number. The right end points of the n subintervals are $\frac{1}{n}, \frac{2}{n}, \frac{3}{n}, \ldots, \frac{n}{n}$. Imagine that for each subinterval there is a disc (a right circular cylinder) having the length of this subinterval as altitude. The x-axis passes through the center of the circular base of the disc. Moreover, the radius of the disc is the number $f(x)$, where x is the right end point of the subinterval and where

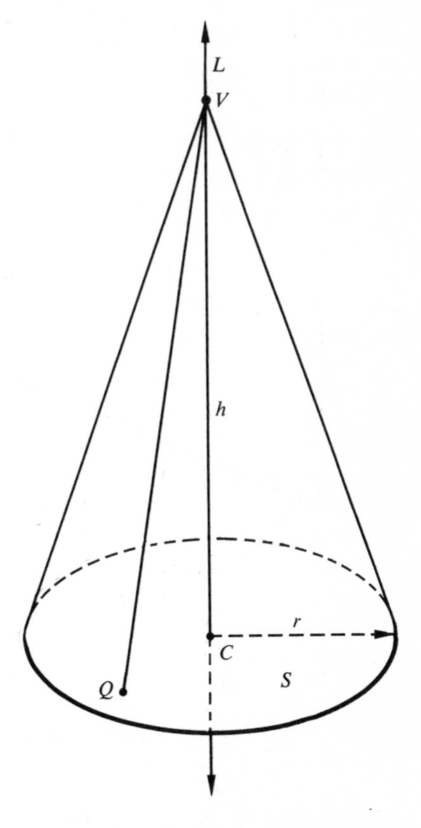

Figure 50. Right Circular Cone

$f(x) = rx$ is the function describing the line which passes through points $(0, 0)$ and $(1, r)$. (See Figure 51.)

The volume of each disc is the product of the altitude $\left(\text{i.e., } \frac{1}{n}\right)$ and the area of the circular base of that disc. The area of the circular base of each disc is πR^2, where $R = f(x) = rx$, with x the right end point of the corresponding subinterval.

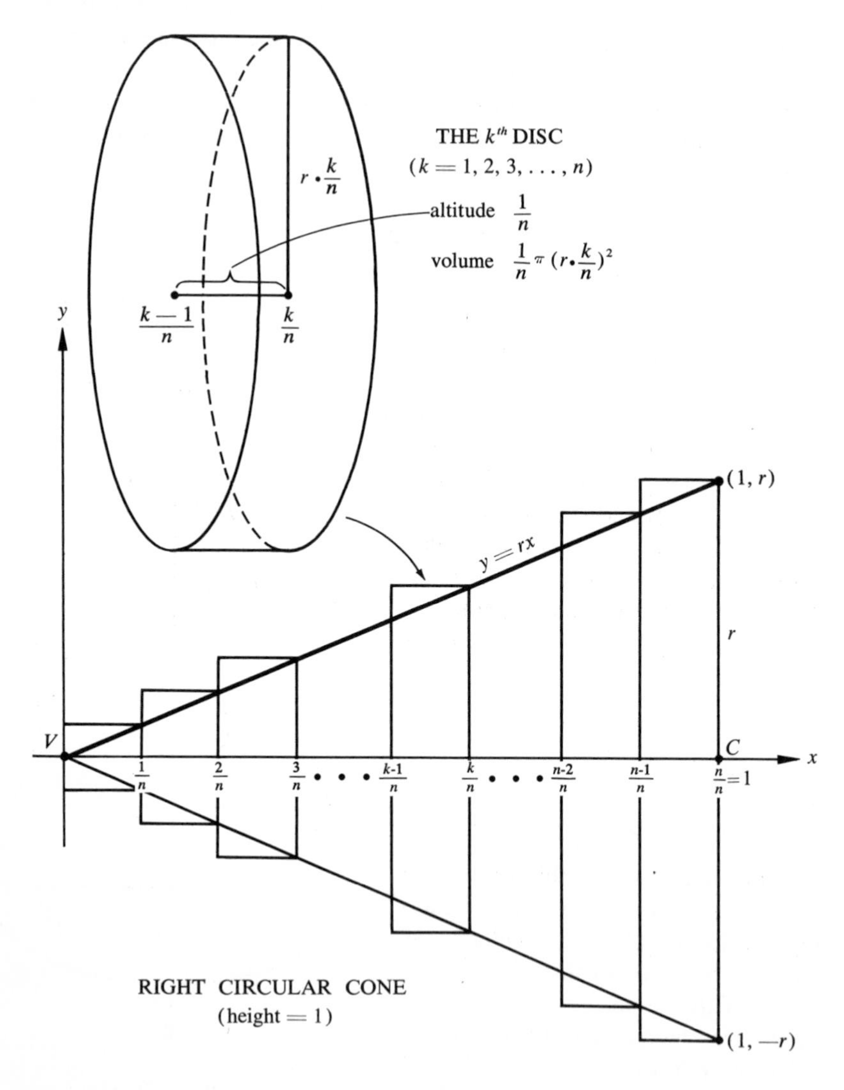

Figure 51

The drawing in Figure 52 represents the case in which n is 4. Each of the four discs has an altitude of $\frac{1}{4}$, and the respective base radii are $r(\frac{1}{4})$, $r(\frac{2}{4})$, $r(\frac{3}{4})$, and $r(\frac{4}{4}) = r$. The respective areas of these bases are $\pi[r(\frac{1}{4})]^2$, $\pi[r(\frac{2}{4})]^2$, $\pi[r(\frac{3}{4})]^2$, and $\pi[r(\frac{4}{4})]^2$, or $\frac{\pi r^2}{4^2}(1^2)$, $\frac{\pi r^2}{4^2}(2^2)$, $\frac{\pi r^2}{4^2}(3^2)$, and $\frac{\pi r^2}{4^2}(4^2)$. The sum of the *volumes* of these four discs is

$$\frac{1}{4} \cdot \frac{\pi r^2}{4^2}[1^2 + 2^2 + 3^2 + 4^2] = \frac{\pi r^2}{4^3}[30] = \frac{15}{32}\pi r^2 \text{ cubic units.}$$

For n discs, where n is any natural number, the sum of the volumes of the n discs will be

$$V_n = \frac{1}{n}\pi\left[r\left(\frac{1}{n}\right)\right]^2 + \frac{1}{n}\pi\left[r\left(\frac{2}{n}\right)\right]^2 + \frac{1}{n}\pi\left[r\left(\frac{3}{n}\right)\right]^2 + \cdots + \frac{1}{n}\pi\left[r\left(\frac{n}{n}\right)\right]^2$$

$$= \frac{\pi r^2}{n^3}[1^2 + 2^2 + 3^2 + \cdots + n^2]$$

$$= \frac{\pi r^2}{n^3}\left[\frac{n(n+1)(2n+1)}{6}\right] = \frac{\pi r^2}{6}\left(1 + \frac{1}{n}\right)\left(2 + \frac{1}{n}\right).$$

Since $\{V_n\}$ converges to $\frac{1}{3}\pi r^2$, this number is regarded as the volume of a right circular cone having $h = 1$ and r any positive real number.

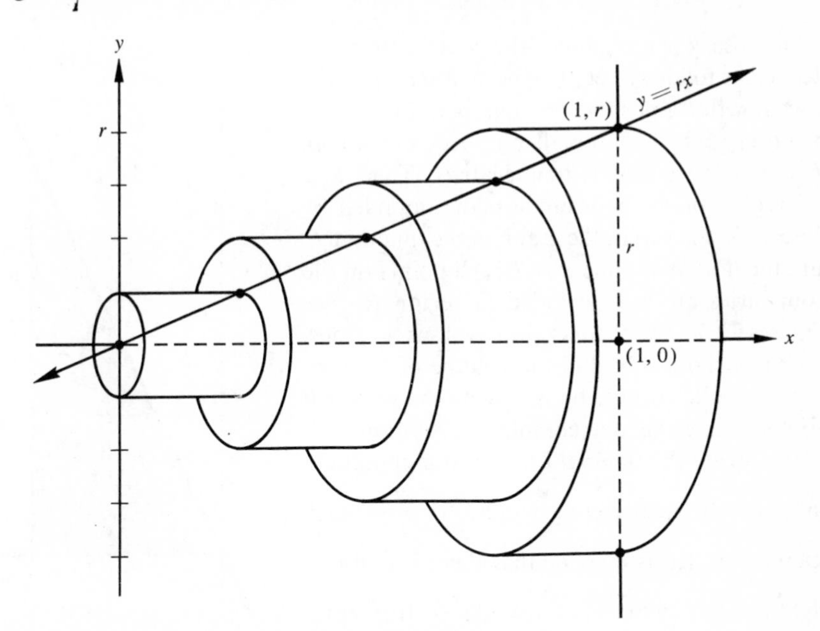

Figure 52

EXERCISES

1. (a) Define a sequence whose limit is the volume of a right circular cone having radius of 3 and altitude $h = 2$. [Hint: Divide the interval [0, 2] on the x-axis into n congruent subintervals. The points of subdivision will then be 0, $\frac{2}{n}, \frac{4}{n}, \frac{6}{n}, \ldots, \frac{2n}{n}$.]
(b) Find the limit of this sequence.

2. (a) Define a sequence whose limit is the volume of a right circular cone having radius r and altitude h. [Hint: Divide the interval $[0, h]$ on the x-axis into n congruent subintervals. The points of subdivision will then be 0, $\frac{h}{n}, \frac{2h}{n}, \frac{3h}{n}, \ldots, \frac{nh}{n}$.]
(b) Find the limit of this sequence.

3. Use the formula developed in Problem 2 to find the volume of a right circular cone having
(a) radius = 5 inches, altitude = 4 inches.
(b) radius = 5 inches, altitude = 8 inches.

4. Volumes of Solids of Revolution

Suppose R is a region of the coordinate plane defined as follows: Let $[0, h]$ be an interval on the x-axis, with h a positive real number. Let f be a function such that, for all numbers x in $[0, h]$, $f(x)$ is defined and is nonnegative. Then R is the region of the coordinate plane bounded by the x-axis, the y-axis, the graph of the function f, and the horizontal line $x = h$. All points on the boundaries are also included in region R. See Figures 53, 54, and 55 for examples of such regions.

Our purpose is to find the volume of the geometrical solid formed by revolving the region R around the x-axis one complete revolution.

We divide the interval $[0, h]$ into n congruent subintervals, each having width $\frac{h}{n}$. For each subinterval, let us imagine that there is a disc (a right circular cylinder) of altitude $\frac{h}{n}$. If x represents the *right* end point of the subinterval (we

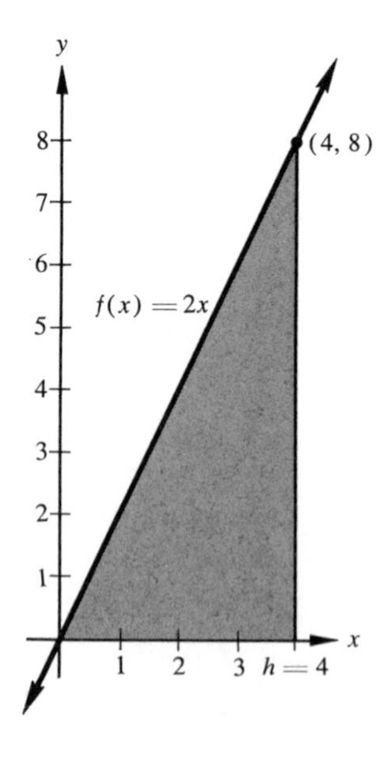

Figure 53

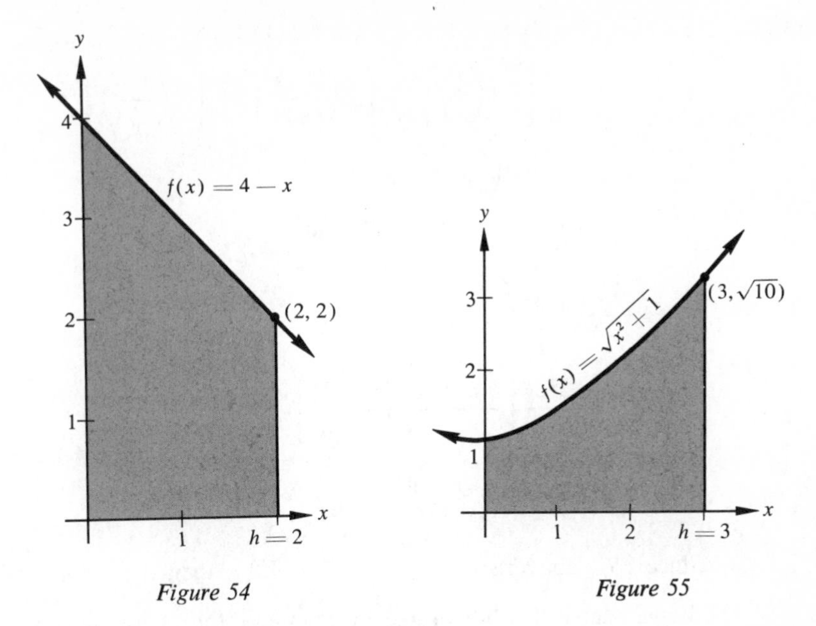

Figure 54

Figure 55

could use the left end point if we wished) then the number $f(x)$ is the radius of the disc. For $x = \frac{h}{n}, \frac{2h}{n}, \frac{3h}{n}, \ldots, \frac{nh}{n}$ the corresponding discs will have radii of

$$f\left(\frac{h}{n}\right), f\left(\frac{2h}{n}\right), f\left(\frac{3h}{n}\right), \ldots, f\left(\frac{nh}{n}\right).$$

The volume of each disc is the product of the altitude and the area of the circular base, which is $\pi[f(x)]^2$. The volumes of the first three discs, for example, are

$$\frac{h}{n}\pi\left[f\left(\frac{h}{n}\right)\right]^2, \quad \frac{h}{n}\pi\left[f\left(\frac{2h}{n}\right)\right]^2, \quad \text{and } \frac{h}{n}\pi\left[f\left(\frac{3h}{n}\right)\right]^2.$$

It is now easy to define a sequence of partial sums whose limit is the volume of the geometrical solid formed by revolving the region R around the x-axis one complete revolution. The sequence has the general term

$$S_n = \frac{h}{n}\pi\left[\left[f\left(\frac{h}{n}\right)\right]^2 + \left[f\left(\frac{2h}{n}\right)\right]^2 + \left[f\left(\frac{3h}{n}\right)\right]^2 + \cdots + \left[f\left(\frac{nh}{n}\right)\right]^2\right]$$

Now that we have this general expression for S_n, we need only to specify a given function and a given value for h in order to make $\{S_n\}$ a particular sequence whose limit can be determined. We shall now do this.

Example **(1)** Let $f(x) = 2x$ and $h = 4$. (See Figure 53 on page 136.)

$$\text{Then } S_n = \frac{4\pi}{n}\left[\left[2\left(\frac{4}{n}\right)\right]^2 + \left[2\left(\frac{2\cdot 4}{n}\right)\right]^2 + \left[2\left(\frac{3\cdot 4}{n}\right)\right]^2 + \cdots + \left[2\left(\frac{n\cdot 4}{n}\right)\right]^2\right]$$

$$= \frac{4\pi}{n}\left[\frac{8^2\cdot 1^2}{n^2} + \frac{8^2\cdot 2^2}{n^2} + \frac{8^2\cdot 3^2}{n^2} + \cdots + \frac{8^2\cdot n^2}{n^2}\right]$$

$$= \frac{4\cdot 8^2\pi}{n^3}[1^2 + 2^2 + 3^2 + \cdots + n^2]$$

$$= \frac{256\pi}{n^3}\left[\frac{n(n+1)(2n+1)}{6}\right]$$

$$= \frac{256}{6}\pi\left(1 + \frac{1}{n}\right)\left(2 + \frac{1}{n}\right).$$

Since $\{S_n\}$ converges to $\frac{256}{6}\pi\cdot 1\cdot 2 = 85\frac{1}{3}\pi$, then the volume of the geometrical solid formed by revolving the region shown in Figure 53 is $85\frac{1}{3}\pi$ cubic units.

REMARK: In the preceding example, with $f(x) = 2x$ and $h = 4$, the solid formed is a right circular cone with altitude 4 and base radius of 8. The method presented in this section is a generalization of the method presented earlier for the volume of a right circular cone.

Example **(2)** Let $f(x) = 4 - x$ and $h = 2$. (See Figure 54 on page 137 and Figure 56 opposite.)

$$\text{Then } S_n = \frac{2\pi}{n}\left[\left[f\left(\frac{2}{n}\right)\right]^2 + \left[f\left(\frac{4}{n}\right)\right]^2 + \left[f\left(\frac{6}{n}\right)\right]^2 + \cdots + \left[f\left(\frac{2n}{n}\right)\right]^2\right]$$

$$= \frac{2\pi}{n}\left[\left[4 - \frac{2}{n}\right]^2 + \left[4 - \frac{4}{n}\right]^2 + \left[4 - \frac{6}{n}\right]^2 + \cdots + \left[4 - \frac{2n}{n}\right]^2\right]$$

$$= \frac{2\pi}{n}\left[\left(16 - \frac{16}{n} + \frac{4}{n^2}\right) + \left(16 - \frac{32}{n} + \frac{16}{n^2}\right) + \left(16 - \frac{48}{n} + \frac{36}{n^2}\right) + \cdots + \left(16 - \frac{16n}{n} + \frac{(2n)^2}{n^2}\right)\right]$$

$$= \frac{2\pi}{n}\left[(16 + 16 + 16 + \cdots + 16) - \left(\frac{16}{n} + \frac{32}{n} + \frac{48}{n} + \cdots + \frac{16n}{n}\right) + \left(\frac{4}{n^2} + \frac{16}{n^2} + \frac{36}{n^2} + \cdots + \frac{(2n)^2}{n^2}\right)\right]$$

$$= \frac{2\pi}{n}\left[16n - \frac{16}{n}(1 + 2 + 3 + \cdots + n) + \frac{4}{n^2}(1^2 + 2^2 + 3^2 + \cdots + n^2)\right]$$

$$= \frac{2\pi}{n}\left[16n - \frac{16}{n}\left(\frac{n+1}{2}\cdot n\right) + \frac{4}{n^2}\left(\frac{n(n+1)(2n+1)}{6}\right)\right]$$

$$= 32\pi - 16\pi\left(1 + \frac{1}{n}\right) + \frac{4}{3}\pi\left(1 + \frac{1}{n}\right)\left(2 + \frac{1}{n}\right).$$

So $\{S_n\}$ converges to $32\pi - 16\pi + \frac{8}{3}\pi = 18\frac{2}{3}\pi$ cubic units.

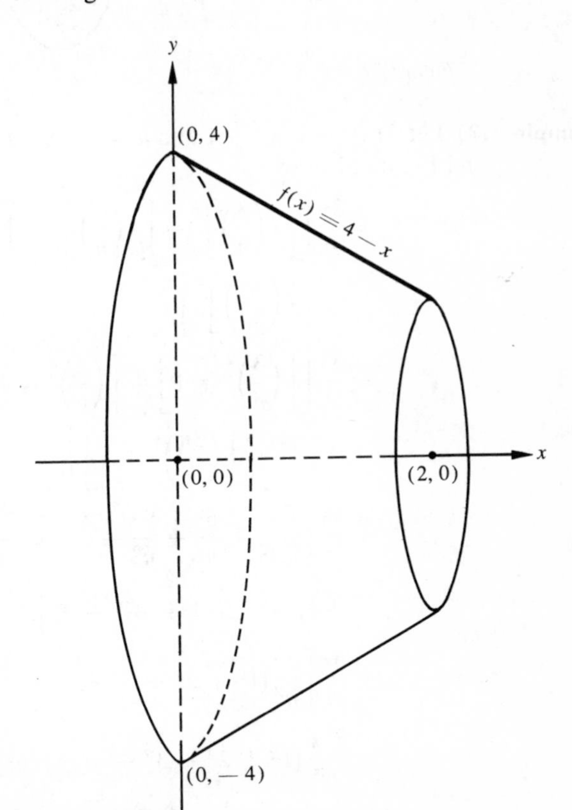

Figure 56

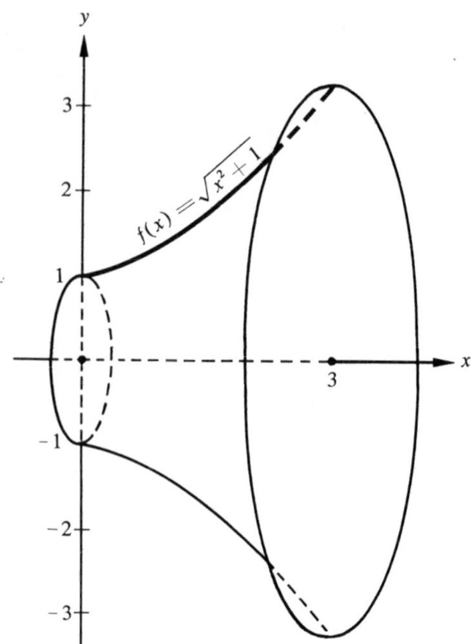

Figure 57

Example **(3)** Let $f(x) = \sqrt{x^2 + 1}$ and $h = 3$. (See Figure 55 on page 137 and Figure 57 above.)

$$\text{Then } S_n = \frac{3\pi}{n}\left[\left[f\left(\frac{3}{n}\right)\right]^2 + \left[f\left(\frac{6}{n}\right)\right]^2 + \left[f\left(\frac{9}{n}\right)\right]^2 + \cdots + \left[f\left(\frac{3n}{n}\right)\right]^2\right]$$

$$= \frac{3\pi}{n}\left[\left[\left(\frac{3}{n}\right)^2 + 1\right] + \left[\left(\frac{6}{n}\right)^2 + 1\right] + \left[\left(\frac{9}{n}\right)^2 + 1\right] + \cdots + \left[\left(\frac{3n}{n}\right)^2 + 1\right]\right]$$

$$= \frac{3\pi}{n}\left[\left[\frac{3^2}{n^2} + \frac{6^2}{n^2} + \frac{9^2}{n^2} + \cdots + \frac{3^2n^2}{n^2}\right] + [1 + 1 + 1 + \cdots + 1]\right]$$

$$= \frac{3\pi}{n}\left[\frac{3^2}{n^2}[1^2 + 2^2 + 3^2 + \cdots + n^2] + n\right]$$

$$= \frac{3^3\pi}{n^3}[1^2 + 2^2 + 3^2 + \cdots + n^2] + 3\pi$$

$$= \frac{27\pi}{n^3}\left[\frac{n(n+1)(2n+1)}{6}\right] + 3\pi$$

$$= \frac{27\pi}{6}\left(1+\frac{1}{n}\right)\left(2+\frac{1}{n}\right) + 3\pi.$$

So $\{S_n\}$ converges to $\dfrac{27\pi}{6}\cdot 1\cdot 2 + 3\pi = 12\pi$.

Thus the volume of the solid obtained by revolving the region in Figure 55 around the x-axis is 12π cubic units. This geometrical solid is shown in Figure 57.

EXERCISES

1. Find the volume of the geometrical solid obtained by revolving each of the given regions around the x-axis.
   (a) The region for which $f(x) = \sqrt{x}$ and $h = 1$.
   (b) The region for which $f(x) = x + 2$ and $h = 3$.
   (c) The region for which $f(x) = \sqrt{x+2}$ and $h = 2$.
   (d) The region shown in Figure 58.

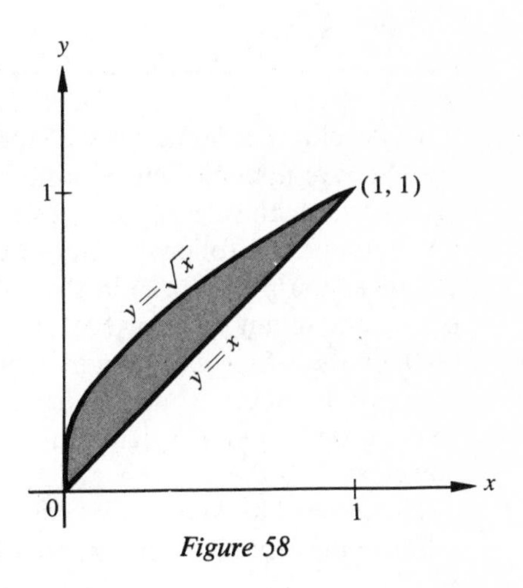

Figure 58

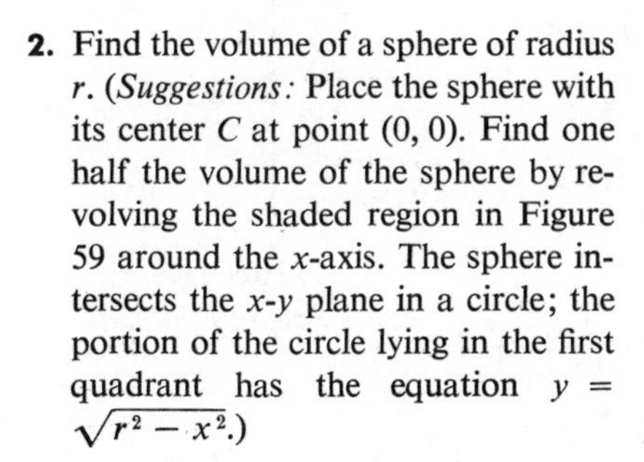

2. Find the volume of a sphere of radius r. (*Suggestions:* Place the sphere with its center C at point $(0, 0)$. Find one half the volume of the sphere by revolving the shaded region in Figure 59 around the x-axis. The sphere intersects the x-y plane in a circle; the portion of the circle lying in the first quadrant has the equation $y = \sqrt{r^2 - x^2}$.)

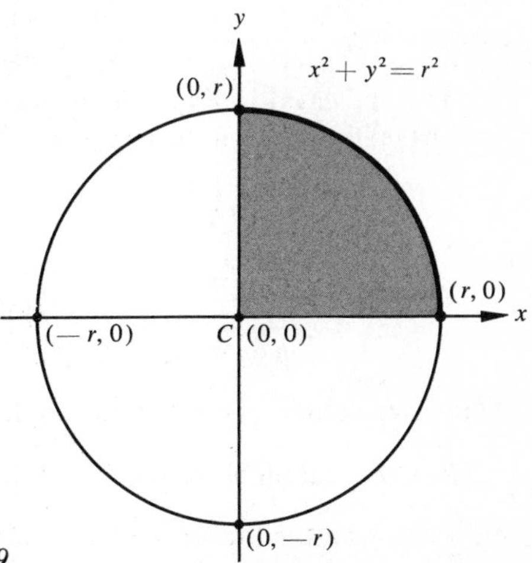

Figure 59

5. Volume of a Pyramid

DEFINITION

Let set S consist of a square and all points in the interior of the square. Let line L be the straight line which is perpendicular to the plane containing set S and which is equidistant from the four vertices of the square. Let C be the point of intersection of set S and line L. Let V be any specified point on line L except C. Then the union of all line segments QV, where Q is any point in S, is called a *right square pyramid* with base S and vertex V. The altitude of the right square pyramid is the length of line segment VC. (See Figure 60.)

To develop a formula for the volume of a right square pyramid having altitude h and square base with side length of s, we position the pyramid as follows: The vertex V is placed at the point $(0, 0)$ in the coordinate plane, and point C is placed at the point $(h, 0)$ on the x-axis, as shown in Figure 61 on the opposite page. Thus the line L referred to in the definition of a right square pyramid coincides with the x-axis.

The interval $[0, h]$ on the x-axis is divided into n congruent subintervals, each of width $\frac{h}{n}$. The points of subdivision of $[0, h]$ are therefore $0, \frac{h}{n}, \frac{2h}{n}, \frac{3h}{n}, \ldots, \frac{nh}{n}$.

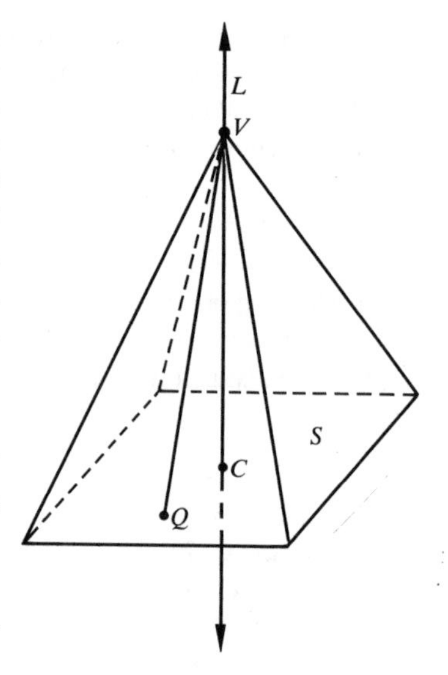

Figure 60
Right Square Pyramid

As is indicated in Figure 61 on the opposite page, the slope of the line joining points $V = (0, 0)$ and $A = \left(h, \frac{s}{2}\right)$ is

$$\frac{\frac{s}{2} - 0}{h - 0} = \frac{s}{2h}.$$

This means that this line is the graph of the function $f(x) = \frac{s}{2h}x$.

Imagine that for each subinterval there is a right square prism whose height is $\frac{h}{n}$ and whose base is a square with each side twice the f-value of the right

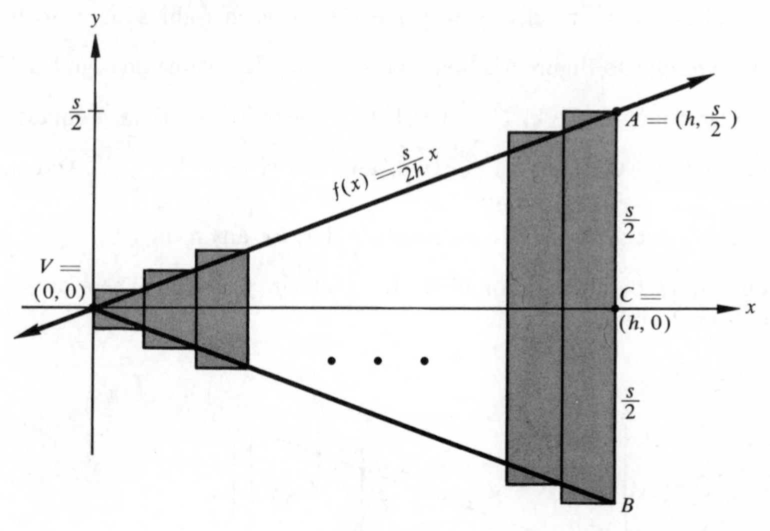

Figure 61

end point, where $f(x) = \frac{s}{2h}x$. That is, the length of each side of the base is equal to $2\left[\frac{s}{2h}x\right]$, or $\frac{s}{h}x$, where x is the right end point of the corresponding subinterval. Figure 62 illustrates the case in which $n = 4$.

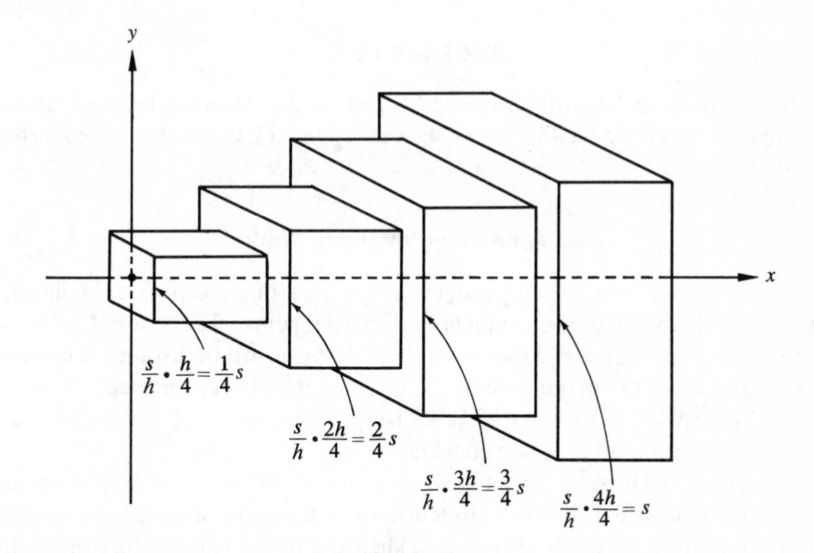

Figure 62

To understand more clearly why the side of each right square prism is $\frac{s}{h}x$, we can refer to Figure 63. Let d be an arbitrary x-value on segment VC. By similar triangles ($\triangle VPQ$ and $\triangle VAB$), the ratio of PQ to d equals the ratio of AB to VC. That is, $\frac{PQ}{d} = \frac{s}{h}$, and this gives $PQ = \frac{s}{h}d$. And since we chose d arbitrarily on VC, we conclude that for any x on VC, $PQ = \frac{s}{h}x$.

You should be able to complete tl is problem yourself. Therefore, it is given as an exercise.

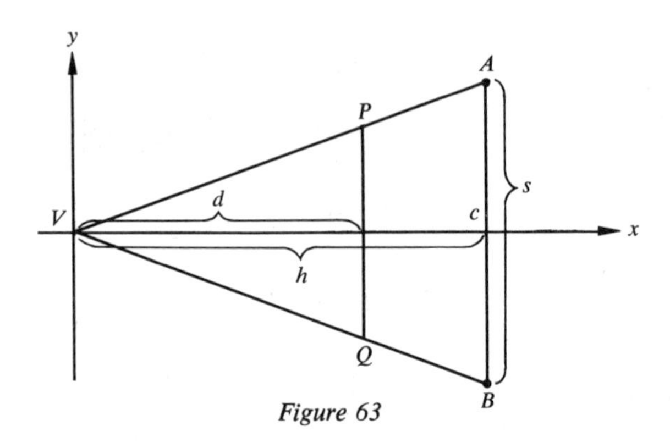

Figure 63

EXERCISE

Devise a sequence of partial sums whose limit is the volume of the right square pyramid, as described in the preceding discussion. Find this limit, and hence the volume.

6. A Problem Involving Work

The concept of work is encountered in physics and can provide a simple, yet powerful, application of sequences of partial sums. The amount of work done when a constant force moves a certain body a certain distance is defined as the product of the magnitude of the force by the distance moved. That is, $W = FD$, provided the force F is constant. However, let us consider an event in which the force is not constant.

According to Hooke's law, the force required to stretch a helical spring is proportional to the distance stretched. For example, suppose the natural length of a given spring is 10 inches. Then the force required to stretch the spring to a total length of $(10 + x)$ inches is kx, where k is a constant. For our present problem, assume $k = 2$. Then the force required to stretch the spring x inches beyond its normal length is given by the formula $F(x) = 2x$ pounds. Now we would like to know how much *work* is required to stretch the spring a certain distance beyond its normal length of 10 inches, say a distance of 3 inches. Then $F(x) = 2x$ and $0 \leq x \leq 3$.

Since the force is variable rather than constant, the formula $W = FD$ is not appropriate. The work involved in stretching the spring from a total length of 10 to 11 inches should certainly be less than the work involved in stretching the spring from 12 to 13 inches. Let us specify, by use of the limit concept, the mathematical meaning which can be associated with this physical problem.

We place the spring on the x-axis, in the interval $[-10, 0]$, and let the left end remain fixed at the point corresponding to -10. Then if we stretch the spring along the x-axis, the positive x value reached by the right end represents the length the spring has been stretched beyond its normal length. We are interested in the amount of work involved in stretching the spring to $x = 3$.

In the interval $[0, 3]$ the force varies from 0 to 6 pounds. So certainly W is less than $6 \cdot 3 = 18$ inch-lbs. That is, $0 < W < 18$.

Divide the interval $[0, 3]$ into n congruent subintervals, each of length $3/n$. Let W_1 be the amount of work done in stretching the spring from 0 to $3/n$, let W_2 be the amount of work done in stretching the spring from $3/n$ to $6/n$, and in general, for k any natural number from 1 to n, let W_k be the amount of work done in stretching the spring in the kth subinterval.

In the subinterval $\left[0, \frac{3}{n}\right]$ the force varies from 0 to $\frac{3}{n} \cdot 2$ (recalling that $F(x) = 2x$), so that $0 \leq W_1 \leq \left(\frac{3}{n}\right)\left(\frac{3}{n} \cdot 2\right)$. In the second subinterval, the force varies from $\frac{3}{n} \cdot 2$ to $\frac{6}{n} \cdot 2$, so that $\left(\frac{3}{n}\right)\left(\frac{3}{n} \cdot 2\right) \leq W_2 \leq \left(\frac{3}{n}\right)\left(\frac{6}{n} \cdot 2\right)$. Similarly, $\left(\frac{3}{n}\right)\left(\frac{6}{n} \cdot 2\right) \leq W_3 \leq \left(\frac{3}{n}\right)\left(\frac{9}{n} \cdot 2\right)$, etc.

We devise a sequence $\{U_n\}$, where

$$
\begin{aligned}
U_n &= \left(\frac{3}{n}\right)\left(\frac{3}{n} \cdot 2\right) + \left(\frac{3}{n}\right)\left(\frac{6}{n} \cdot 2\right) + \left(\frac{3}{n}\right)\left(\frac{9}{n} \cdot 2\right) + \cdots + \left(\frac{3}{n}\right)\left(\frac{3n}{n} \cdot 2\right) \\
&= \frac{18}{n^2}[1 + 2 + 3 + \cdots + n] = \frac{18}{n^2}\left[\frac{n+1}{2} \cdot n\right] = 9\left(1 + \frac{1}{n}\right).
\end{aligned}
$$

The limit of $\{U_n\}$ is 9, and we can define this limit to be the actual amount of work required to stretch the spring three inches beyond its natural length. With the theory of limits, problems such as this one can be solved!

7. Similarities Among the Area, Volume, and Work Problems

In this chapter we have developed certain applications of sequences of partial sums to area, volume and work problems. It is quite likely that participating in these developments has enhanced your understanding of the importance of the idea of a limit. Now that solutions to these problems involving area, volume and work have unfolded, it should be enlightening for us to take notice of the common properties shared by these solutions.

In general, we proceed as follows: For h some positive real number and for g some function which is nonnegative everywhere in the interval $[0, h]$, we divide the interval $[0, h]$ into n congruent subintervals, where n is any natural number. For each subinterval we imagine an element (often called a *typical element*) of some kind; for example,

(a) The element may be a rectangle, if we wish to find the area under the curve $y = g(x)$.

(b) The element may be a disc if we wish to find the volume of a geometrical solid generated by revolving a planar region around the x-axis.

(c) The element may be the amount of work required to stretch a spring along that subinterval.

There are many other kinds of *typical elements* besides the ones mentioned in this chapter, for we have selected only a very few types to demonstrate some of the applications of sequences of partial sums.

In all of these situations we form a sequence of partial sums $\{S_n\}$ whose general term is

$$S_n = \frac{h}{n} g\left(\frac{h}{n}\right) + \frac{h}{n} g\left(\frac{2h}{n}\right) + \frac{h}{n} g\left(\frac{3h}{n}\right) + \cdots + \frac{h}{n} g\left(\frac{nh}{n}\right)$$
$$= \frac{h}{n}\left[g\left(\frac{h}{n}\right) + g\left(\frac{2h}{n}\right) + g\left(\frac{3h}{n}\right) + \cdots + g\left(\frac{nh}{n}\right)\right].$$

The limit of $\{S_n\}$ (provided this limit exists) is the desired area, volume, or amount of work spent, as the situation calls for.

Examples **(1)** To find the area of the shaded region in Figure 64 we consider a typical element to be a rectangle with width $2/n$ and height $g(x) = 3x^2$, where x is the right end point of the subinterval.

Then $$S_n = \frac{2}{n}\left[g\left(\frac{2}{n}\right) + g\left(\frac{4}{n}\right) + g\left(\frac{6}{n}\right) + \cdots + g\left(\frac{2n}{n}\right)\right]$$
$$= \frac{2}{n}\left[3\left(\frac{2}{n}\right)^2 + 3\left(\frac{4}{n}\right)^2 + 3\left(\frac{6}{n}\right)^2 + \cdots + 3\left(\frac{2n}{n}\right)^2\right]$$
$$= \frac{2}{n} \cdot \frac{12}{n^2}[1^2 + 2^2 + 3^2 + \cdots + n^2]$$

$$= \frac{24}{n^3}\left[\frac{n(n+1)(2n+1)}{6}\right] = 4\left(1+\frac{1}{n}\right)\left(2+\frac{1}{n}\right), \text{ so}$$

$\{S_n\} \to 8$. Therefore, the area of the shaded region in Figure 64 is 8 square units.

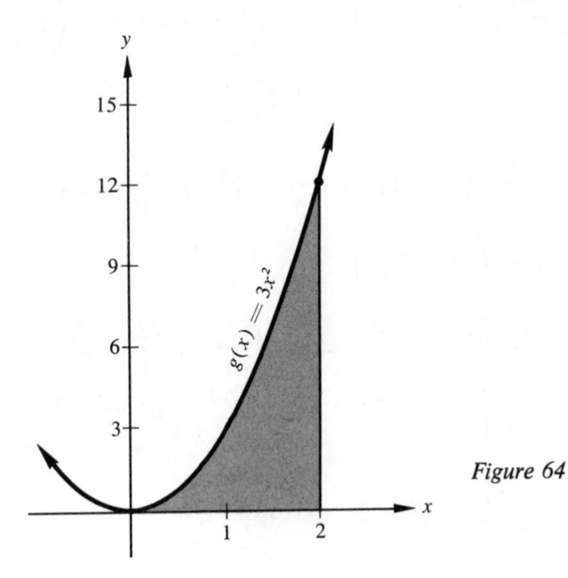

Figure 64

(2) To find the volume of the geometrical solid formed by revolving the shaded region in Figure 65 around the x-axis, we consider a typical element to be a disc whose altitude is $2/n$ and whose radius is $f(x) = 3x + 2$, where x is the right end point of the subinterval. For this value of x, $g(x) = \pi[f(x)]^2 = \pi[3x+2]^2$ is the area of the circular base of the disc.

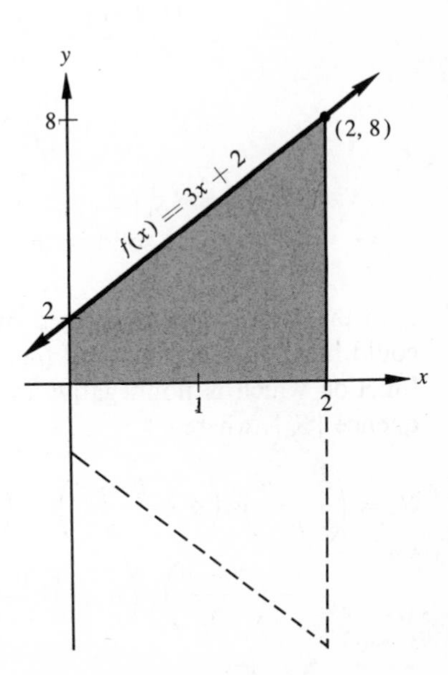

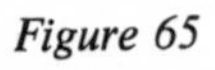

Figure 65

The volume of the solid is the limit of the sequence $\{S_n\}$, where

$$S_n = \frac{2}{n}\left[g\left(\frac{2}{n}\right) + g\left(\frac{4}{n}\right) + g\left(\frac{6}{n}\right) + \cdots + g\left(\frac{2n}{n}\right)\right]$$

$$= \frac{2}{n}\left[\pi\left(3\cdot\frac{2}{n} + 2\right)^2 + \pi\left(3\cdot\frac{4}{n} + 2\right)^2 + \pi\left(3\cdot\frac{6}{n} + 2\right)^2 + \cdots + \pi\left(3\cdot\frac{2n}{n} + 2\right)^2\right]$$

$$= \frac{2\pi}{n}\left[\left(\frac{6}{n} + 2\right)^2 + \left(\frac{12}{n} + 2\right)^2 + \left(\frac{18}{n} + 2\right)^2 + \cdots + \left(\frac{6n}{n} + 2\right)^2\right]$$

$$= \frac{2\pi}{n}\left[\left(\frac{6^2}{n^2} + \frac{24}{n} + 4\right) + \left(\frac{6^2\cdot 4}{n^2} + \frac{48}{n} + 4\right) + \left(\frac{6^2\cdot 9}{n^2} + \frac{72}{n} + 4\right) + \cdots + \left(\frac{6^2\cdot n^2}{n^2} + \frac{24n}{n} + 4\right)\right].$$

$$= \frac{2\pi}{n}\left[\frac{6^2}{n^2}(1^2 + 2^2 + 3^2 + \cdots + n^2) + \frac{24}{n}(1 + 2 + 3 + \cdots + n) + 4n\right]$$

$$= \frac{72\pi}{n^3}\left[\frac{n(n+1)(2n+1)}{6}\right] + \frac{48\pi}{n^2}\left[\frac{n+1}{2}\cdot n\right] + 8\pi$$

$$= 12\pi\left(1 + \frac{1}{n}\right)\left(2 + \frac{1}{n}\right) + 24\pi\left(1 + \frac{1}{n}\right) + 8\pi.$$

Since $\{S_n\} \to 24\pi + 24\pi + 8\pi = 56\pi$, the volume of the geometrical solid is 56π cubic units.

In this chapter we have used only the interval $[0, h]$. More generally, we could have used any interval $[a, b]$ on the x-axis. Still assuming that g is a function which is nonnegative everywhere in $[a, b]$, we would form the sequence $\{S_n\}$, where

$$S_n = \left(\frac{b-a}{n}\right)g\left(a + \frac{b-a}{n}\right) + \left(\frac{b-a}{n}\right)g\left(a + 2\cdot\frac{b-a}{n}\right) + \left(\frac{b-a}{n}\right)g\left(a + 3\cdot\frac{b-a}{n}\right) + \cdots + \left(\frac{b-a}{n}\right)g\left(a + n\cdot\frac{b-a}{n}\right)$$

$$= \left(\frac{b-a}{n}\right)\left[g\left(a + \frac{b-a}{n}\right) + g\left(a + 2\cdot\frac{b-a}{n}\right) + g\left(a + 3\cdot\frac{b-a}{n}\right) + \cdots + g\left(a + n\cdot\frac{b-a}{n}\right)\right].$$

To avoid lengthy calculations, we have used the case when $a = 0$, obtaining

$$S_n = \frac{b}{n}\left[g\left(\frac{b}{n}\right) + g\left(2\cdot\frac{b}{n}\right) + g\left(3\cdot\frac{b}{n}\right) + \cdots + g\left(n\cdot\frac{b}{n}\right)\right].$$

In courses in calculus, you will have many very interesting opportunities for linking the ideas developed in this chapter with techniques of calculus, particularly those techniques which come under the heading "integration."

7 • *Infinite Series*

1. Introduction

Definitions Relating to Infinite Series

DEFINITIONS

If $\{a_n\}$ is a sequence, the expression $a_1 + a_2 + a_3 + \cdots + a_n + \cdots$ is called an *infinite series*, or simply a *series*. Corresponding to every infinite series is the *sequence of partial sums* $\{S_n\} = \{a_1 + a_2 + a_3 + \cdots + a_n\}$.

An infinite series $a_1 + a_2 + a_3 + \cdots + a_n + \cdots$ is said to be *convergent* if $\{S_n\}$ is convergent, and *divergent* if $\{S_n\}$ is divergent. If $\{S_n\} \to L$, it is customary to write $a_1 + a_2 + a_3 + \cdots + a_n + \cdots = L$ and to refer to L as the *sum* of the infinite series.

Use of the word *sum* does not imply that any term of $\{S_n\}$ is equal to the number L. Nor does it imply that all of the numbers a_i $(i = 1, 2, 3, \ldots)$ have been added. This would clearly be impossible. It is simply another name for the limit of the sequence of partial sums. It is very important to bear in mind that the following statements are merely different ways of expressing the same idea:

"The sum of $a_1 + a_2 + a_3 + \cdots + a_n + \cdots$ is L."

"$a_1 + a_2 + a_3 + \cdots + a_n + \cdots = L$."

"$\{a_1 + a_2 + a_3 + \cdots + a_n\} \to L$."

"$a_1 + a_2 + a_3 + \cdots + a_n + \cdots$ converges to L."

Purposes of this Chapter

The topic "infinite series" is not really separate from that of "sequences," for *infinite series* is an expression commonly used in place of the expression *sequence of partial sums.* We see from the above definitions that convergence for an infinite series is defined in terms of convergence for a sequence: For every neighborhood of the sum of the series, there must be a particular term of the corresponding sequence of partial sums such that this term and all succeeding terms of the sequence of partial sums are in the given neighborhood.

In showing convergence of infinite series, we will not use the definition directly. Instead, we will develop certain theorems for determining convergence or divergence, just as we did in Chapters 3 and 5 for sequences. In fact, we will relate many of the theorems of those two chapters to the techniques essential for handling infinite series.

Please keep in mind that tests for convergence or divergence of series usually do not specify the actual sum. Most of these tests indicate only whether or not the sum does exist. These tests are nevertheless powerful and valuable, for very often, merely knowing for certain that a given series converges or diverges is sufficient for the purposes on hand. With the use of high-speed computers the sum of almost any convergent series can be calculated to any specified number of decimal places in a very short time.

This chapter is an extension of the preceding chapter, which was concerned with sequences of partial sums. For each series of the preceding chapter, a general term could be found which represented the nth partial sum. This term is useful in determining the sum of a series (whenever the sum exists). For example, in finding a certain area, we might develop a sequence of partial sums such as $\left\{9\left(1+\frac{1}{n}\right)\left(2+\frac{1}{n}\right)\right\}$, whose convergence to 18 would be immediately apparent. However, there are a great many kinds of series for which a general term has not been found, and is not expected to be found. It is series such as these which we will study in this chapter.

The topic "infinite series" is a very broad topic and, one might say, a very distinguished topic. It has captured the imaginations of the most gifted mathematicians, on the one hand, and of interested students, on the other hand. Over the centuries, interest in this topic has directly and indirectly led to great discoveries, and as you study this topic, the way is open for you to relive some of these discoveries.

In making our present survey of infinite series, we shall barely skim the surface, not only because the topic is so broad, but also because a full treatment would presuppose a knowledge of calculus. In particular, the proofs of several theorems must be delayed until you have studied more mathematics.

2. Families of Series

Trigonometric Series

Sometimes a *family* of infinite series is described by a single general term, by using the letter "x" (or some other letter) to denote any member of a specified set of real numbers. For example, if x is any real number, then

$$\sin x = x - \frac{x^3}{3!} + \frac{x^5}{5!} - \frac{x^7}{7!} + \cdots + \frac{(-1)^{n-1}x^{2n-1}}{(2n-1)!} + \cdots,$$

and $$\cos x = 1 - \frac{x^2}{2!} + \frac{x^4}{4!} - \frac{x^6}{6!} + \cdots + \frac{(-1)^{n-1}x^{2n-2}}{(2n-2)!} + \cdots.$$

In particular, for $x = 1$ (measured in radians, not degrees),

$$\sin 1 = 1 - \frac{1}{3!} + \frac{1}{5!} - \frac{1}{7!} + \cdots + \frac{(-1)^{n-1}}{(2n-1)!} + \cdots.$$

The first few terms of the corresponding sequence of partial sums are 1.00000, 0.83333, 0.84166, and 0.84140 (rounded off to five decimal places). The value of sin 1, as listed in a five-place table, is 0.84147. For $x = 1$ and other values of x which are small in absolute value, both the sine series and the cosine series converge quite rapidly. For values of x which are large in absolute value, these two series still converge, though not quite so rapidly.

The expression "$\text{Tan}^{-1} x$" (read "the inverse tangent of x"), where x is any real number, is used to mean the real number y in the interval $\langle -\pi/2, \pi/2 \rangle$ such that the tangent of y is x. For example, $\text{Tan}^{-1} 1 = \frac{\pi}{4}$, since $\tan \frac{\pi}{4} = 1$. (The expression "Arctan x" is often used in place of $\text{Tan}^{-1} x$.)

If $|x| \leq 1$, then

$$\text{Tan}^{-1} x = x - \tfrac{1}{3}x^3 + \tfrac{1}{5}x^5 - \tfrac{1}{7}x^7 + \cdots + \frac{(-1)^{n-1}x^{2n-1}}{2n-1} + \cdots,$$

and if $|x| > 1$, then

$$\text{Tan}^{-1} x = \frac{\pi}{2} - \frac{1}{x} + \frac{1}{3x^3} - \frac{1}{5x^5} + \frac{1}{7x^7} + \cdots + \frac{(-1)^n}{(2n-1)x^{2n-1}} + \cdots.$$

In the special case when $x = 1$, the first of these two inverse tangent series gives an expansion which can be used to compute an approximation for π: Since $\tan\left(\frac{\pi}{4}\right) = 1$, $\text{Tan}^{-1}(1) = \frac{\pi}{4}$. So

$$\frac{\pi}{4} = 1 - \tfrac{1}{3} + \tfrac{1}{5} - \tfrac{1}{7} + \cdots + \frac{(-1)^{n-1}}{2n-1} + \cdots,$$

and this gives

$$\pi = 4\left[1 - \tfrac{1}{3} + \tfrac{1}{5} - \tfrac{1}{7} + \cdots + \frac{(-1)^{n-1}}{2n-1} + \cdots\right].$$

For $n = 8$ we have $4[1 - \frac{1}{3} + \frac{1}{5} - \frac{1}{7} + \frac{1}{9} - \frac{1}{11} + \frac{1}{13} - \frac{1}{15}] = 4(0.754268) = 3.017072$. By using larger and larger values for n, we can eventually obtain the approximation 3.1416 which we often use for π.

Trigonometric series are invaluable for performing various calculations. Almost all entries in tables of trigonometric functions are approximations of sums of various series.

Geometric Series

Geometric sequences of partial sums were studied in the preceding chapter. It was noted that $\{S_n\} = \{a + ar + ar^2 + \cdots + ar^{n-1}\} = \left\{\frac{a(1 - r^n)}{1 - r}\right\}$ $= \left\{\frac{a}{1 - r} - \frac{ar^n}{1 - r}\right\}$, and that this sequence converges to $\frac{a}{1 - r}$ if and only if $0 < |r| < 1$ and diverges if and only if $|r| \geq 1$. Therefore, the geometric series $a + ar + ar^2 + ar^3 + \cdots + ar^{n-1} + \cdots$ has the sum $\frac{a}{1 - r}$ if and only if $0 < |r| < 1$, and it diverges if and only if $|r| \geq 1$.

The above general form for a geometric series describes a family of geometric series: for each value of a (any real number except 0) and for each value of r we obtain a particular member of the family.

The Number e and Natural Logarithms

Suppose \$10,000 is invested at 4% interest for one year. If interest is compounded annually, the investment will be worth \$10,400 at the end of the year. If the interest is compounded semiannually, the investment will be worth $\$10{,}000(1.02)^2 = \$10{,}404$. If the interest is compounded quarterly, the investment will be worth yet slightly more at the end of the year, namely $\$10{,}000(1.01)^4 = \$10{,}406.04$. We observe that the more often the interest is compounded, the greater the return on the investment.

Let us examine a similar situation. Suppose \$1.00 is invested for one year at the unusual interest rate of 100%! Let n be the number of interest periods per year, and let a_n be the amount of the investment at the end of the year. If interest is compounded semiannually, for example, then we have $a_2 = (1 + \frac{1}{2})^2 = \$2.25$. If interest is compounded quarterly, then $a_4 = (1 + \frac{1}{4})^4 = \$2.44$. In general, for n any natural number,

$$a_n = \left(1 + \frac{1}{n}\right)^n.$$

We might wonder how much the investment would be worth if we were permitted a very large number of interest periods, say 3650. That is, we might like to know how much the investment would be worth if we compounded the interest ten times per day. We need not actually calculate

$$a_{3650} = \left(1 + \frac{1}{3650}\right)^{3650},$$

since the sequence $\{a_n\} = \left\{\left(1 + \frac{1}{n}\right)^n\right\}$ is a well-known sequence, which can be shown to be everywhere increasing and bounded above by 3. (Details of these two facts are quite long and hence are omitted.) This means that the sequence converges. Its limit, to eight decimal places, is 2.71828183.

Therefore, no matter how often the interest on \$1.00 is compounded, the investment at the end of the year will never be worth more than \$2.72. The maximum return would represent the situation in which the interest is compounded instantaneously, at every instant, for a year.

This discussion leads us to a very special number called e, which is the limit of the sequence $\{a_n\} = \left\{\left(1 + \frac{1}{n}\right)^n\right\}$. The example involving interest at 100% is only one particular application of this special sequence.

In courses prior to calculus, the number 10 is generally used as a base for logarithms. However, for many purposes the number e is the best choice for a base. When you study calculus, you will then be able to understand the reasons for the superiority of e as a base for logarithms. One reason for the use of e as a base is the existence of an infinite series whose sum is e. More generally, there exists a series expansion for e^x, where x is any real number:

$$e^x = 1 + x + \frac{x^2}{2!} + \frac{x^3}{3!} + \frac{x^4}{4!} + \cdots + \frac{x^n}{n!} + \cdots.$$

In the special case when $x = 1$,

$$e = 1 + 1 + \frac{1}{2!} + \frac{1}{3!} + \frac{1}{4!} + \cdots + \frac{1}{n!} + \cdots,$$

and it can be shown that the sum of this series is the same as the limit of the sequence $\{a_n\} = \left\{\left(1 + \frac{1}{n}\right)^n\right\}$ described previously.

Regarding the series for e^x, the following information is apparent:

When $x = 0$, $e^x = 1$. When $x > 1$, $e^x > e$.

When $x = 1$, $e^x = e \approx 2.718$. When $x < 0$, $0 < e^x < 1$.

Logarithms to the base e are known as natural logarithms. It should be remembered that a logarithm is an exponent to a given base. For example, if $e^x = y$, then $\log_e y = x$; the base is e and x is the logarithm.

The series

$$e^x = 1 + x + \frac{x^2}{2!} + \frac{x^3}{3!} + \cdots + \frac{x^n}{n!} + \cdots$$

can be used to compile natural logarithmic values. The series

$$\log_e (1 + x) = x - \frac{x^2}{2} + \frac{x^3}{3} - \frac{x^4}{4} + \cdots + \frac{(-1)^{n-1}x^n}{n} + \cdots,$$

which converges for $-1 < x \leq 1$, can also be used. In particular, for $x = 1$ we have

$$\log_e 2 = 1 - \tfrac{1}{2} + \tfrac{1}{3} - \tfrac{1}{4} + \cdots + \frac{(-1)^{n-1}}{n} + \cdots = 0.6931 \text{ (approximately).}$$

This series for $\log_e x$ is limited, since it converges only for $-1 < x \leq 1$. However, this series can be used to form another series by which the logarithm of any positive real number can be found. To show this we proceed as follows:

(1) First note that

$$\log_e (1 + x) = x - \frac{x^2}{2} + \frac{x^3}{3} - \frac{x^4}{4} + \cdots + \frac{(-1)^{n-1}x^n}{n} + \cdots,$$

and

$$\log_e (1 - x) = -x - \frac{x^2}{2} - \frac{x^3}{3} - \frac{x^4}{4} - \cdots - \frac{x^n}{n} - \cdots.$$

(2) Since in general $\log \frac{a}{b} = \log a - \log b$, then

$$\begin{aligned} \log_e \left[\frac{1 + x}{1 - x}\right] &= \log_e (1 + x) - \log_e (1 - x) \\ &= 2x + \frac{2x^3}{3} + \frac{2x^5}{5} + \cdots + \frac{2x^{2n-1}}{2n - 1} + \cdots \\ &= 2\left[x + \frac{x^3}{3} + \frac{x^5}{5} + \cdots + \frac{x^{2n-1}}{2n - 1} + \cdots\right] \end{aligned}$$

for $-1 < x < 1$.

(3) If c is any positive real number, then $\frac{1 + x}{1 - x} = c \leftrightarrow x = \frac{c - 1}{c + 1}$.

For example, if $c = 3$, then $x = \frac{1}{2}$ and

$$\begin{aligned} \log_e 3 = \log_e \frac{1 + \frac{1}{2}}{1 - \frac{1}{2}} &= 2\left[\frac{1}{2} + \frac{(\frac{1}{2})^3}{3} + \frac{(\frac{1}{2})^5}{5} + \cdots + \frac{(\frac{1}{2})^{2n-1}}{2n - 1} + \cdots\right] \\ &= 2\left[\frac{1}{2} + \frac{1}{2^3 \cdot 3} + \frac{1}{2^5 \cdot 5} + \cdots\right] \\ &= 1.0986 \text{ (approximately).} \end{aligned}$$

An Historical Note Concerning Logarithms

Let a_n be the number of primes among the first n natural numbers. For example, a_{10} is 4, since there are four primes (namely 2, 3, 5, and 7) among the natural numbers 1, 2, 3, $\cdots$, 9, 10. The first five terms of this sequence

$\{a_n\}$ are 0, 1, 2, 2, 3. The famous mathematician Gauss (1777–1855), whom many people refer to as The Prince of Mathematicians, conjectured that the ratio $\frac{a_n}{n}$ is approximately equal to $\frac{1}{\log n}$ and that this approximation improves as n increases. In fact, the sequence $\left\{\frac{a_n}{n}(\log n)\right\}$ converges to 1. It was considered quite amazing that the logarithmic function so accurately describes the average distribution of the prime numbers, since in other respects the two concepts, logarithmic functions and prime number theory, seemed so unrelated at that time. Gauss's conjecture — called the Prime Number Theorem — was not proved completely until almost a hundred years after it was made.

EXERCISES

1. Find the sum, whenever it exists, of

(a) $1 + x + x^2 + x^3 + \cdots + x^{n-1} + \cdots$, for $x = -\frac{1}{2}$.

(b) $.17 + .0017 + .000017 + \cdots + 17(.01)^n + \cdots$.

(c) $1.04 + 1.04^2 + 1.04^3 + \cdots + 1.04^n + \cdots$.

2. Find a series whose sum is

(a) $\frac{.7}{1 - .1} = \frac{7}{9}$ (c) $\frac{3}{11}$

(b) $\frac{.2}{1 - .7} = \frac{2}{3}$ (d) $\frac{2}{7}$

3. Show that for $|x| > 1$,

$$\text{Tan}^{-1} x = \frac{\pi}{2} - \frac{1}{x} + \frac{1}{3x^3} - \frac{1}{5x^5} + \frac{1}{7x^7} - \cdots + \frac{(-1)^n}{(2n-1)x^{2n-1}} + \cdots,$$

using the series given in the text for $\text{Tan}^{-1} x$ for $|x| \leq 1$ and the identity $\text{Tan}^{-1} x + \text{Tan}^{-1}\left(\frac{1}{x}\right) = \frac{\pi}{2}$.

4. In the two series given in the text for $\text{Tan}^{-1} x$, what series are obtained by replacing x with $\tan x$?

5. Sketch a graph of the function

(a) $f(x) = e^x$ domain: $\langle -\infty, \infty \rangle$
range: $\langle 0, \infty \rangle$

(b) $g(x) = \log_e x$ domain: $\langle 0, \infty \rangle$
range: $\langle -\infty, \infty \rangle$

6. Find a series whose sum is $\log_e 9$, and find the first two terms of the corresponding sequence of partial sums, $\{S_n\}$.

7. Dividing the number 1 by the polynomial $1 - x$ gives

$$\frac{1}{1-x} = 1 + x + x^2 + x^3 + \cdots + x^{n-1} + \cdots, (x \neq 1).$$

For $x = -1$ the series $1 - 1 + 1 - 1 + \cdots$ is obtained, and seventeenth-century mathematicians were amazed to find that its sum is the number $\frac{1}{1-x} = \frac{1}{1-(-1)} = \frac{1}{2}$, since the terms of this series are 1, 0, 1, 0, 1, 0, $\cdots$. Some mathematicians suggested that $\frac{1}{2}$ really was the limit of this series, since $\frac{1}{2}$ is the average of the two numbers 0 and 1. Can *you* explain this paradox?

3. Extending the Ideas of Chapter 3

Let us consider only those infinite series $a_1 + a_2 + a_3 + \cdots + a_n + \cdots$ for which all terms of the sequence $\{a_n\}$ are positive. This means that each corresponding sequence of partial sums $\{S_n\} = \{a_1 + a_2 + a_3 + \cdots + a_n\}$ is a sequence which is everywhere increasing. In Chapter 3 we dealt in detail with increasing sequences, so our present discussion is really an extension of the major ideas developed in Chapter 3. In dealing with infinite series we should certainly take advantage of pertinent principles regarding sequences which have already been established. After all, a series, as depicted by its corresponding sequence of partial sums, is itself a special kind of sequence.

Pertinent Principles Recalled from Chapter 3

It will be helpful at this point to recall the main ideas associated with increasing sequences which were developed in Chapter 3:

(1) If $\{a_n\}$ is everywhere increasing, then $\{a_n\}$ is convergent if and only if it is bounded above. (This statement is a summary of several theorems of Chapter 3.)

(2) **Theorem 3-1 (Comparison Principle).** Let $\{a_n\}$ be a sequence which increases without bound, and let $\{b_n\}$ be a sequence which is everywhere increasing. If there is a particular natural number M such that, for all natural numbers $n \geq M$, $b_n \geq a_n$, then $\{b_n\}$ also increases without bound.

(3) **Theorem 3-2.** Suppose $\{a_n\}$ is everywhere increasing and suppose that $\{b_n\}$ increases without bound *and* is a subsequence of $\{a_n\}$. Then $\{a_n\}$ increases without bound.

(4) **Theorem 3-3.** Two special families of sequences which increase without bound are

(a) sequences of the form $\{n^p\}$, where p is any positive real number.
(b) sequences of the form $\{p^n\}$, where p is any real number > 1.

Applying the Principles From Chapter 3

The first of the four principles gives us immediately the following theorem:

THEOREM 7-1

If all terms of $\{a_n\}$ are positive, then the series $a_1 + a_2 + a_3 + \cdots + a_n + \cdots$ is convergent if and only if the corresponding sequence of partial sums is bounded above.

Next, suppose the series $a_1 + a_2 + a_3 + \cdots + a_n + \cdots$ (all $a_n > 0$) diverges, and suppose $b_1 + b_2 + b_3 + \cdots + b_n + \cdots$ (all $b_n > 0$) is a series such that $b_n \geq a_n$ for all natural numbers n. Let

$$\{A_n\} = \{a_1 + a_2 + a_3 + \cdots + a_n\}$$

and $\{B_n\} = \{b_1 + b_2 + b_3 + \cdots + b_n\}$. Then $A_n \leq B_n$ for all natural numbers n. Thus by Theorem 3-1, $\{B_n\}$ increases without bound, and hence diverges. That is, the series $b_1 + b_2 + b_3 + \cdots + b_n + \cdots$ diverges.

On the other hand, suppose $a_1 + a_2 + a_3 + \cdots + a_n + \cdots$ converges and $b_n \leq a_n$ for all natural numbers n. This means that the series $b_1 + b_2 + b_3 + \cdots + b_n + \cdots$ converges, for if this series were to diverge, then so would the series $a_1 + a_2 + a_3 + \cdots + a_n + \cdots$, according to the conclusion of the preceding paragraph. We have now proved the next theorem.

THEOREM 7-2

Comparison Test. Assume $\{a_n\}$ and $\{b_n\}$ are two sequences such that, for all natural numbers n, $0 < a_n \leq b_n$. Then

(i) If $a_1 + a_2 + a_3 + \cdots + a_n + \cdots$ diverges,
$b_1 + b_2 + b_3 + \cdots + b_n + \cdots$ also diverges.

(ii) If $b_1 + b_2 + b_3 + \cdots + b_n + \cdots$ converges,
$a_1 + a_2 + a_3 + \cdots + a_n + \cdots$ also converges.

Examples of applications of the Comparison Test will be presented after two other theorems have been introduced.

Directly from Theorem 3-2 we obtain the next theorem.

THEOREM 7-3

If $\{a_n\}$ increases without bound and is a subsequence of the sequence of partial sums $\{b_1 + b_2 + b_3 + \cdots + b_n\}$ (all $b_n > 0$), then the series $b_1 + b_2 + b_3 + \cdots + b_n + \cdots$ increases without bound, and hence diverges.

The next theorem is not a direct consequence of the principles recalled from Chapter 3, but its introduction at this time is essential.

THEOREM 7-4(a)

If a series $a_1 + a_2 + a_3 + \cdots + a_n + \cdots$, with all a_n positive, converges, then the sequence $\{a_n\}$ converges to 0.

Proof

Let $\{S_n\} = \{a_1 + a_2 + a_3 + \cdots + a_n\} \to L$. Then for $\langle L - E, L + E\rangle$ any neighborhood of L, there is a particular natural number M' such that $S_{M'}$ and all terms S_n with $n \geq M'$ are in the neighborhood $\langle L - E, L + E\rangle$. Then all terms of $\{a_n\}$ with $n > M'$ are less than or equal to $|S_{M'} - L|$. For suppose there is a natural number $j > M'$ such that $a_j > |S_{M'} - L|$. Then, by Remark 1 on page 112, $S_{M'}$ is in $\langle L - a_j, L + a_j\rangle$, so $L - a_j < S_{M'}$, or

$$S_{M'} + a_j > L.$$

But since all terms of $\{a_n\}$ are positive,

$$S_j = S_{M'} + a_{M'+1} + \cdots + a_j \geq S_{M'} + a_j,$$

so $S_j > L$. This is impossible because L is the least upper bound of $\{S_n\}$ by Theorem 3-4 (since $\{S_n\}$ is everywhere increasing and bounded above by L).

Now let $M = M' + 1$. Then all terms of $\{a_n\}$ with $n \geq M$ are less than or equal to $|S_{M'} - L|$, and therefore less than E, since $|S_{M'} - L| < E$. This means that all terms of $\{a_n\}$ with $n \geq M$ are in the neighborhood $\langle 0 - E, 0 + E\rangle$, and thus that $\{a_n\} \to 0$. (See Figure 66.)

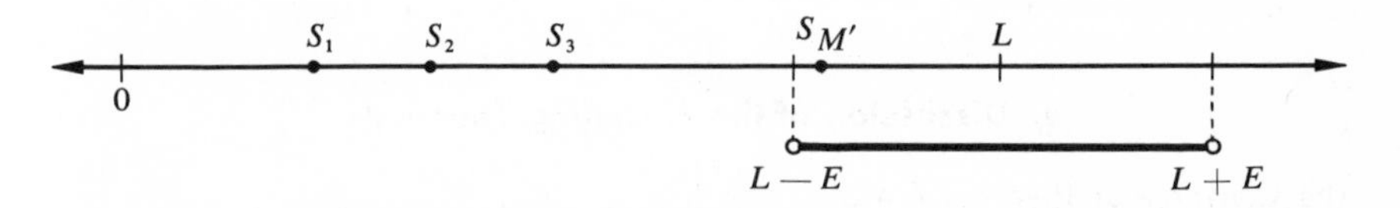

Figure 66

The following theorem, Theorem 7-4(b), is simply a restatement of Theorem 7-4(a) and therefore needs no proof.

THEOREM 7-4(b)

If $\{a_n\}$, a positive term sequence, does not converge to 0, then the series $a_1 + a_2 + a_3 + \cdots + a_n + \cdots$ diverges. (In other words, if $\{a_n\}$ diverges, or converges to some number other than 0, then the series diverges.)

Examples **(1)** $1 + 3 + 5 + \cdots + (2n - 1) + \cdots$ diverges, since $\{2n - 1\}$ diverges.

(2) $\frac{1}{2} + \frac{2}{3} + \frac{3}{4} + \cdots + \dfrac{n}{n+1} + \cdots$ diverges, because $\left\{\dfrac{n}{n+1}\right\} \to 1.$

(3) If $\{a_n\} = \{c\}$ is any constant sequence, then $\{a_n\}$ converges to c. Thus for $c > 0$ the series $c + c + c + \cdots + c + \cdots$ diverges.

Finally, as a result of Theorems 3-3 and 7-4(b), we have the following theorem.

THEOREM 7-5

The series $a_1 + a_2 + a_3 + \cdots + a_n + \cdots$, with a_n positive for all n, diverges if $\{a_n\}$ is of either of the forms

(a) $\{n^p\}$, where p is a real number > 0,
(b) $\{p^n\}$, where p is a real number > 1.

For example, each of the following series diverges:

$$\sqrt{1} + \sqrt{2} + \sqrt{3} + \cdots + \sqrt{n} + \cdots$$

$$1^2 + 2^2 + 3^2 + \cdots + n^2 + \cdots$$

$$3 + 3^2 + 3^3 + \cdots + 3^n + \cdots$$

4. Discussion of the Preceding Theorems

The Converse of Theorem 7-4(a)

The converse of Theorem 7-4(a) can be stated as follows:

If $\{a_n\} \to 0$, where $\{a_n\}$ is a positive term sequence, then $a_1 + a_2 + a_3 + \cdots + a_n + \cdots$ converges.

Is this a true statement?

For example, since $\{a_n\} = \left\{\dfrac{1}{n}\right\} \to 0$, does the series

$$1 + \tfrac{1}{2} + \tfrac{1}{3} + \tfrac{1}{4} + \cdots + \frac{1}{n} + \cdots$$

converge? Let us answer this question on the basis of the following two examples.

Examples **(1)** Let $\{t_n\}$ be defined as follows:

The first term is 1;
the next two terms are each $\frac{1}{2}$;
the next three terms are each $\frac{1}{3}$;
the next four terms are each $\frac{1}{4}$; and
each subsequent set of n terms is $\frac{1}{n}\cdot$

Since $\{a_n\} = \left\{\frac{1}{n}\right\} \to 0$, $\{t_n\}$ also converges to 0, although it converges more slowly than does $\{a_n\} = \left\{\frac{1}{n}\right\}\cdot$

If $T_n = t_1 + t_2 + t_3 + \cdots + t_n$, then $T_1 = 1, T_3 = 2, T_6 = 3, T_{10} = 4$, and in general,

$$T_{\frac{n+1}{2}\cdot n} = n.$$

Since $\{T_n\}$ contains the sequence $\{b_n\} = \{n\}$ as a subsequence, then by Theorem 7-3 the series $t_1 + t_2 + t_3 + \cdots + t_n + \cdots$ diverges! Here is an example of a sequence $\{t_n\}$ which converges to 0, but whose corresponding infinite series is divergent. Thus the converse of Theorem 7-4(a) is false.

The next example also shows that the converse of Theorem 7-4(a) is false.

(2) Let $\{v_n\}$ be a sequence described as follows:

The first term is 1;
the second term is $\frac{1}{2}$;
the next *two* terms are each $\frac{1}{4}$;
the next *four* terms are each $\frac{1}{8}$;
the next *eight* terms are each $\frac{1}{16}$; and,
in general, the next 2^n terms are each $\frac{1}{2^{n+1}}\cdot$

For $\{V_n\}$, the corresponding sequence of partial sums, we have $V_1 = 1$, $V_2 = 1\frac{1}{2}$, $V_4 = 2$, $V_8 = 2\frac{1}{2}$, etc. In general,

$$V_{2^{n-1}} = 1 + \frac{n-1}{2} = \frac{n+1}{2}\cdot$$

Since $\{V_n\}$ contains the sequence $\{a_n\} = \left\{\frac{n+1}{2}\right\}$ as a subsequence, and since $\{a_n\}$ increases without bound, the series $v_1 + v_2 + v_3 + \cdots + v_n + \cdots$ diverges by Theorem 7-3.

Multiplication of a Series by a Constant

Chapter 5 included the theorem "If $\{a_n\} \to A$ and $\{b_n\} \to B$, then $\{a_n b_n\} \to AB$."

Suppose we are given a constant sequence, say $\{a_n\} = \{A\}$, and a sequence of partial sums $\{B_n\}$ with limit B. Then $\{a_n B_n\} = \{AB_n\}$ is a sequence of partial sums whose limit is AB. Notice that the first three terms of $\{B_n\}$ are b_1, $b_1 + b_2$, and $b_1 + b_2 + b_3$, while the first three terms of $\{AB_n\}$ are Ab_1, $Ab_1 + Ab_2$, and $Ab_1 + Ab_2 + Ab_3$.

For example, if $\{a_n\} = \{3\}$ and $\{B_n\} = \{1 + \frac{1}{4} + \frac{1}{16} + \cdots + (\frac{1}{4})^{n-1}\} \to \frac{4}{3}$, then $\{a_n B_n\} = \{3 + \frac{3}{4} + \frac{3}{16} + \cdots + 3(\frac{1}{4})^{n-1}\} \to 3 \cdot \frac{4}{3} = 4$.

The following theorem can now be stated.

THEOREM 7-6

If $\{c\}$ is a constant sequence ($c > 0$) and $a_1 + a_2 + a_3 + \cdots + a_n + \cdots$, with a_n positive for all natural numbers n, is a series whose sum is A, then the sum of the series $ca_1 + ca_2 + ca_3 + \cdots + ca_n + \cdots$ is cA.

Example Convergence of $5 + \frac{5}{2} + \frac{5}{3} + \frac{5}{4} + \cdots + \frac{5}{n} + \cdots$ depends upon convergence of $1 + \frac{1}{2} + \frac{1}{3} + \cdots + \frac{1}{n} + \cdots$, and vice versa. Since we know that the latter series diverges, we conclude that the former also diverges.

Uses of the Comparison Test (Theorem 7-2)

The Comparison Test is often useful for showing convergence or divergence. As a basis for comparison we can, of course, use any series whose convergence or divergence has been proved in the theorems, the examples, or the exercises. The series which have been discussed to date are listed in summary:

(1) The series $1 + \frac{1}{2} + \frac{1}{3} + \cdots + \frac{1}{n} + \cdots$ diverges.

(2) A geometric series $a + ar + ar^2 + \cdots + ar^{n-1} + \cdots$ converges to $\frac{a}{1-r}$ if and only if $0 < |r| < 1$.

(3) Theorem 7-5. The series $a_1 + a_2 + a_3 + \cdots + a_n + \cdots$, with a_n positive for all natural numbers n, diverges if $\{a_n\}$ is of either of the forms

(a) $\{n^p\}$, where p is a real number > 0,
(b) $\{p^n\}$, where p is a real number > 1.

Examples **(1)** Let $a_n = \frac{1}{n}$ and $b_n = \frac{1}{\sqrt{2n-1}}$.

Then
$$\frac{1}{\sqrt{2n-1}} \geq \frac{1}{n} \leftrightarrow \sqrt{2n-1} \leq n$$
$$\leftrightarrow 2n - 1 \leq n^2$$
$$\leftrightarrow 2 - \frac{1}{n} \leq n,$$ which is true for all natural numbers n.

Thus $b_n \geq a_n$ for all natural numbers n, so the series $b_1 + b_2 + b_3 + \cdots + b_n + \cdots$ diverges by comparison with the series $1 + \frac{1}{2} + \frac{1}{3} + \cdots + \frac{1}{n} + \cdots$.

(2) The series $\frac{1}{3} + \frac{1}{8} + \frac{1}{17} + \frac{1}{32} + \cdots + \frac{1}{n^2 + 2^n} + \cdots$ converges by comparison with the geometric series $\frac{1}{2} + \frac{1}{4} + \frac{1}{8} + \frac{1}{16} + \cdots + (\frac{1}{2})^n + \cdots$, since
$$\frac{1}{n^2 + 2^n} \leq (\tfrac{1}{2})^n \leftrightarrow n^2 + 2^n \geq 2^n$$
$$\leftrightarrow n^2 \geq 0,$$ which is true for all natural numbers n.

EXERCISES

1. Show that the series $1 + \frac{1}{2} + \frac{1}{3} + \frac{1}{4} + \cdots + \frac{1}{n} + \cdots$ diverges.

2. Which of the following series diverges by comparison with $1 + \frac{1}{2} + \frac{1}{3} + \frac{1}{4} + \cdots + \frac{1}{n} + \cdots$?

(a) $1 + \frac{1}{2^3} + \frac{1}{3^3} + \frac{1}{4^3} + \cdots + \frac{1}{n^3} + \cdots$

(b) $1 + \frac{1}{\sqrt{2}} + \frac{1}{\sqrt[3]{2}} + \frac{1}{\sqrt[4]{2}} + \cdots + \frac{1}{\sqrt[n]{2}} + \cdots$

3. Let $a_n = \begin{cases} 1 \text{ for } n \text{ odd} \\ 0 \text{ for } n \text{ even.} \end{cases}$

(a) Does $a_1 + a_2 + a_3 + \cdots + a_n + \cdots$ converge? If so, what is its limit?

(b) Let $S_n = \frac{a_1 + a_2 + a_3 + \cdots + a_n}{n}$. Does $\{S_n\}$ converge? If so, what is its limit?

4. Determine which of the following series converge and which ones diverge. Show why.

(a) $1 + \frac{1}{\sqrt[3]{2}} + \frac{1}{\sqrt[3]{3}} + \cdots + \frac{1}{\sqrt[3]{n}} + \cdots$

(b) $5 + 4 + 3 + 2 + \cdots + (6 - n) + \cdots$

(c) $\frac{1}{4} + \frac{2}{7} + \frac{3}{10} + \frac{4}{13} + \cdots + \frac{n}{3n + 1} + \cdots$

(d) $2 + \frac{2}{3} + \frac{2}{7} + \frac{2}{15} + \cdots + \frac{2}{2^n - 1} + \cdots$

(e) $1 + \frac{1}{2} + \frac{4}{13} + \frac{1}{5} + \cdots + \frac{2^n}{3^n - 1} + \cdots$

(f) $10 + \frac{10}{2!} + \frac{10}{3!} + \frac{10}{4!} + \cdots + \frac{10}{n!} + \cdots$

(g) $\frac{1}{12} + \frac{1}{24} + \frac{1}{36} + \cdots + \frac{1}{12n} + \cdots$

(h) $\frac{1}{3} + \frac{1}{18} + \frac{1}{81} + \frac{1}{324} + \cdots + \frac{1}{n3^n} + \cdots$

(i) $\frac{3}{2} + \frac{9}{8} + \frac{27}{24} + \frac{81}{64} + \cdots + \frac{3^n}{n2^n} + \cdots$

(j) $1 + \frac{1}{2} + \frac{1}{3} + \frac{3}{4} + \cdots + \frac{n + (-1)^n(n - 2)}{2n} + \cdots$

(k) $\frac{1}{1 \cdot 2} + \frac{1}{2 \cdot 3} + \frac{1}{3 \cdot 4} + \cdots + \frac{1}{n(n + 1)} + \cdots$

5. Does $1 + \frac{3}{2} + \frac{1}{3} + \frac{3}{4} + \frac{1}{5} + \cdots + \frac{2 + (-1)^n}{n} + \cdots$ converge? If so, what is its limit?

6. Show that for p a real number such that $0 < p < 1$ the series $\frac{1}{1^p} + \frac{1}{2^p} + \frac{1}{3^p} + \cdots + \frac{1}{n^p} + \cdots$ diverges.

5. Alternating Series and the Ratio Test

In this section we no longer deal strictly with positive term sequences.

DEFINITION

If the terms of $\{a_n\}$ are alternately positive and negative, the series $a_1 + a_2 + a_3 + \cdots + a_n + \cdots$ is called an *alternating series.*

THEOREM 7-7

Test for Alternating Series. The alternating series $a_1 + a_2 + a_3 + \cdots + a_n + \cdots$ is convergent if $\{|a_n|\}$ converges to 0 and is everywhere decreasing.

Rather than present a proof of this theorem, it will perhaps be more meaningful to indicate why the theorem is true by considering the series $1 - \frac{1}{2} + \frac{1}{3} - \frac{1}{4} + \cdots + \frac{(-1)^{n-1}}{n} + \cdots$. (If you wish, you can try to construct a proof of the theorem by utilizing the ideas brought forth in this discussion. Remember that you must consider both the case where all odd terms are positive and the case where all odd terms are negative.)

Let $S_n = a_1 + a_2 + a_3 + \cdots + a_n$

$$= 1 + (-\tfrac{1}{2}) + \tfrac{1}{3} + (-\tfrac{1}{4}) + \cdots + \frac{(-1)^{n-1}}{n}$$

$$= 1 - \tfrac{1}{2} + \tfrac{1}{3} - \tfrac{1}{4} + \cdots + \frac{(-1)^{n-1}}{n}.$$

We wish to show that $\{S_n\}$ is convergent.

First, consider the subsequence $\{S_{2n}\}$ having as terms the even terms of $\{S_n\}$. We note that

$$S_2 = 1 - \tfrac{1}{2} = \tfrac{1}{2}$$

$$S_4 = S_2 + (\tfrac{1}{3} - \tfrac{1}{4}) = \tfrac{1}{2} + \tfrac{1}{12} = \tfrac{7}{12}$$

$$S_6 = S_4 + (\tfrac{1}{5} - \tfrac{1}{6}) = \tfrac{7}{12} + \tfrac{1}{30} = \tfrac{37}{60}$$

$$\vdots$$

$$S_{2n} = S_{2n-2} + \left(\frac{1}{2n-1} - \frac{1}{2n}\right), \text{ for all } n \geq 2.$$

Since $\frac{1}{2n-1} - \frac{1}{2n} > 0$ for all n, the sequence $\{S_{2n}\}$ is everywhere increasing, with all terms positive.

Next, by writing

$$S_{2n} = 1 - (\tfrac{1}{2} - \tfrac{1}{3}) - (\tfrac{1}{4} - \tfrac{1}{5}) - \cdots - \left(\frac{1}{2n-2} - \frac{1}{2n-1}\right) - \left(\frac{1}{2n}\right),$$

in which each expression in parentheses represents a positive number, it follows that all terms of $\{S_{2n}\}$ are less than $a_1 = 1$.

Thus the positive-term sequence $\{S_{2n}\}$ is everywhere increasing and is bounded above by $a_1 = 1$. Therefore, by Theorem 3-4, $\{S_{2n}\}$ converges to a number S such that $0 < S \leq 1$.

Next, we note that (1) any odd term S_{2n+1} is the sum of S_{2n} and a_{2n+1}, so that $\{S_{2n+1}\} = \{S_{2n} + a_{2n+1}\} = \{S_{2n}\} + \{a_{2n+1}\}$, and (2) since $\{a_n\} \to 0$, the subsequence $\{a_{2n+1}\}$, with terms $a_3, a_5, a_7, \ldots$, also converges to 0. (This fact is immediate from our alternate definition of convergence in Chapter 2 — since any neighborhood of 0 contains all but a finite number of terms of $\{a_n\}$, it must also contain all but a finite number of terms of the subsequence $\{a_{2n+1}\}$.)

It follows that $\{S_{2n+1}\} = \{S_{2n}\} + \{a_{2n+1}\} \to S + 0 = S$. That is, $\{S_{2n+1}\}$ and $\{S_{2n}\}$ both converge to the same number, S, which means that $\{S_n\} \to S$.

REMARK: In the preceding discussion of

$$\{S_n\} = \left\{1 - \tfrac{1}{2} + \tfrac{1}{3} - \cdots + \frac{(-1)^{n-1}}{n}\right\},$$

it was noted that every even term of $\{S_n\}$ is less than a_1. In addition, we could have shown that every odd term $> S_1$ is less than a_1. In fact, the following generalization can be shown to be true:

Let $a_1 + a_2 + a_3 + \cdots + a_n \cdots$ be an alternating series in which $\{|a_n|\}$ converges to 0 and is everywhere decreasing. Assume that the odd terms of $\{a_n\}$ are positive and the even terms are negative, and let $S_n = a_1 + a_2 + a_3 + \cdots + a_n$. Then for all natural numbers n, $S_n \leq a_1$.

The validity of this generalization can be shown as follows:

$$S_n = a_1 + (a_2 + a_3) + (a_4 + a_5) + \cdots + (a_n) \text{ for } n \text{ even, and}$$

$$S_n = a_1 + (a_2 + a_3) + (a_4 + a_5) + \cdots + (a_{n-1} + a_n) \text{ for } n \text{ odd.}$$

Since all numbers in parentheses are less than or equal to 0, $S_n \leq a_1$ for all natural numbers n.

This generalization will be referred to in the proof of Theorem 7-9.

Example If $\{a_n\} = \left\{(-1)^{n+1}\left[\frac{n}{n^2+1}\right]\right\}$, $\{|a_n|\} = \left\{\frac{n}{n^2+1}\right\} = \left\{\frac{1}{n + \frac{1}{n}}\right\}$ converges to 0. Also,

$$|a_n| \geq |a_{n+1}| \leftrightarrow \frac{n}{n^2+1} \geq \frac{n+1}{(n+1)^2+1}$$

$$\leftrightarrow n^2 + n \geq 1, \text{ which is true for all natural numbers.}$$

Thus $\{|a_n|\}$ is everywhere decreasing. Therefore,

$$\tfrac{1}{2} - \tfrac{2}{5} + \tfrac{3}{10} - \tfrac{4}{17} + \cdots + (-1)^{n+1}\left[\frac{n}{n^2+1}\right] + \cdots$$

converges.

A proof of the next theorem, though not difficult, employs several mathematical ideas encountered in more advanced courses. At this point, a proof would be quite long and involved, and has therefore been omitted.

THEOREM 7-8

The Ratio Test. Given a series $a_1 + a_2 + a_3 + \cdots + a_n + \cdots$, form the sequence

$$\{b_n\} = \left\{\left|\frac{a_{n+1}}{a_n}\right|\right\}.$$

(i) If $\{b_n\}$ converges to a number b, where $0 \leq b < 1$, then the series $a_1 + a_2 + a_3 + \cdots + a_n + \cdots$ converges.

(ii) If $\{b_n\}$ converges to a number $b > 1$, or if $\{b_n\}$ diverges, then the series $a_1 + a_2 + a_3 + \cdots + a_n + \cdots$ diverges.

(iii) If $\{b_n\} \to 1$, the test fails. That is, the series may or may not be convergent; other means of determining convergence or divergence must be employed.

Examples (1) The series $e^x = 1 + x + \frac{x^2}{2!} + \frac{x^3}{3!} + \cdots + \frac{x^n}{n!} + \cdots$ converges for all real numbers x by the Ratio Test, since $\left\{\left|\frac{a_{n+1}}{a_n}\right|\right\}$ $= \left\{\left|\frac{x^{n+1}}{x^n} \cdot \frac{n!}{(n+1)!}\right|\right\} = \left\{\frac{|x|}{n+1}\right\} = \{|x|\}\left\{\frac{1}{n+1}\right\} \to |x| \cdot 0 = 0.$

(2) If $\{a_n\} = \left\{\frac{1}{n}\right\}$, then $\left\{\left|\frac{a_{n+1}}{a_n}\right|\right\} = \left\{\frac{n}{n+1}\right\}$; and if $\{c_n\} = \left\{\frac{(-1)^{n-1}}{n}\right\}$, then $\left\{\left|\frac{c_{n+1}}{c_n}\right|\right\}$ also equals $\left\{\frac{n}{n+1}\right\}$. Since $\left\{\frac{n}{n+1}\right\} \to 1$, the test fails to determine convergence of either

$$a_1 + a_2 + a_3 + \cdots + a_n + \cdots = 1 + \tfrac{1}{2} + \tfrac{1}{3} + \cdots + \frac{1}{n} + \cdots$$

or

$$c_1 + c_2 + c_3 + \cdots + c_n + \cdots =$$
$$1 - \tfrac{1}{2} + \tfrac{1}{3} - \tfrac{1}{4} + \cdots + \frac{(-1)^{n-1}}{n} + \cdots.$$

It may be of interest to note that the first of these two series diverges (as shown earlier) and that the second of these two series converges (as can be shown by the Test for Alternating Series, Theorem 7-7).

(3) The series

$$\text{Tan}^{-1}x = x - \tfrac{1}{3}x^3 + \tfrac{1}{5}x^5 - \tfrac{1}{7}x^7 + \cdots + \frac{(-1)^{n-1}x^{2n-1}}{2n-1} + \cdots$$

converges for $|x| < 1$ and diverges for $|x| > 1$ by the Ratio Test. We shall show this is true as follows:

$$a_n = \frac{(-1)^{n-1}x^{2n-1}}{2n-1} \quad \text{and} \quad \left\{\left|\frac{a_{n+1}}{a_n}\right|\right\} = \left\{x^2\left(\frac{2n-1}{2n+1}\right)\right\} \to x^2.$$

For $|x| < 1, 0 \leq x^2 < 1$, so that $\left\{\left|\frac{a_{n+1}}{a_n}\right|\right\}$ converges to a number < 1, and hence the series converges. For $|x| > 1$, $x^2 > 1$, so $\left\{\left|\frac{a_{n+1}}{a_n}\right|\right\}$ converges to a number > 1, and hence the series diverges. For $|x| = 1$ the Ratio Test fails; in this event, convergence can be shown by the Test for Alternating Series.

REMARKS: To examine an alternating series for convergence, we can proceed in either of two ways:

(i) We can first apply the Test for Alternating Series.

If $\{|a_n|\} \to 0$ and $\{|a_n|\}$ is everywhere decreasing, then the series converges.

If $\{|a_n|\}$ does not converge to 0, then the series diverges.

If $\{|a_n|\} \to 0$, but $\{|a_n|\}$ is not everywhere decreasing, convergence is still uncertain (that is, the test fails). We can then apply the Ratio Test in hopes it might work.

If we are unable to determine whether or not $\{|a_n|\}$ converges to 0 *and* is everywhere decreasing, we can apply the Ratio Test.

(ii) We can first apply the Ratio Test.

If $\left\{\left|\frac{a_{n+1}}{a_n}\right|\right\} \to b$, where $0 \leq b < 1$, the series converges.

If $\left\{\left|\frac{a_{n+1}}{a_n}\right|\right\}$ converges to b, where $b > 1$, or if it diverges, then the series diverges.

If $\left\{\left|\frac{a_{n+1}}{a_n}\right|\right\} \to 1$, the test fails, and we can possibly show convergence by the Test for Alternating Series.

If we are unable to determine convergence or divergence of $\left\{\left|\frac{a_{n+1}}{a_n}\right|\right\}$, we can possibly show convergence by the Test for Alternating Series.

Approximating the Sum of an Alternating Series

For approximating the sum of a convergent alternating series, the following theorem is often very helpful.

THEOREM 7-9

If $a_1 + a_2 + a_3 + \cdots + a_n + \cdots$ is an alternating series such that $\{|a_n|\}$ converges to 0 and is everywhere decreasing, then the error made by using the sum of the first k terms as an approximation for the sum of the series is no greater than $|a_{k+1}|$.

Proof

Let L be the sum of the series $a_1 + a_2 + a_3 + \cdots + a_n + \cdots$ (it exists by Theorem 7-7), and let k be a particular natural number held constant throughout the proof. Then $a_1 + a_2 + a_3 + \cdots + a_n + \cdots = (a_1 + a_2 + a_3 + \cdots + a_k) + R_k = L$, where R_k is the series $a_{k+1} + a_{k+2} + a_{k+3} + \cdots$.

Since R_k is itself an alternating series in which $\{|a_{k+n}|\}$ converges to zero and is everywhere decreasing, then R_k also converges. Let R be the sum of series R_k. Then $L = (a_1 + a_2 + a_3 + \cdots + a_k) + R$.

We wish to show next that $|R| \leq |a_{k+1}|$. Since $|R| = |a_{k+1} + a_{k+2} + a_{k+3} + \cdots|$, then we can show by the following argument that $|R| = |a_{k+1}| - |a_{k+2}| + |a_{k+3}| - \cdots$, an alternating series with all odd terms positive and all even terms negative. If a_{k+1} is positive:

$|R| = \big||a_{k+1}| - |a_{k+2}| + |a_{k+3}| - |a_{k+4}| + \cdots\big|$, by definition of absolute value

$= \big|(|a_{k+1}| - |a_{k+2}|) + (|a_{k+3}| - |a_{k+4}|) + \cdots\big|$

$= (|a_{k+1}| - |a_{k+2}|) + (|a_{+3}| - |a_{k+4}|) + \cdots$, since all expressions in parentheses are positive

$= |a_{k+1}| - |a_{k+2}| + |a_{k+3}| - |a_{k+4}| + \cdots$.

If a_{k+1} is negative:

$|R| = \big|-|a_{k+1}| + |a_{k+2}| - |a_{k+3}| + |a_{k+4}| - \cdots\big|$, by definition of absolute value

$= \big|-(|a_{k+1}| - |a_{k+2}|) - (|a_{k+3}| - |a_{k+4}|) - \cdots\big|$

$= \big|-[(|a_{k+1}| - |a_{k+2}|) + (|a_{k+3}| - |a_{k+4}|) + \cdots]\big|$

$= [(|a_{k+1}| - |a_{k+2}|) + (|a_{k+3}| - |a_{k+4}|) + \cdots]$, by definition of absolute value, since the bracketed expression is positive

$= |a_{k+1}| - |a_{k+2}| + |a_{k+3}| - |a_{k+4}| + \cdots$.

Therefore, from the Remark on page 166, we can conclude that the nth partial sum $(|a_{k+1}| - |a_{k+2}| + |a_{k+3}| - \cdots + (-1)^{n-1}|a_{k+n}|) \leq |a_{k+1}|$ for all natural numbers n, and from this it follows that the sum of $|a_{k+1}| - |a_{k+2}| + |a_{k+3}| - \cdots + (-1)^{n-1}|a_{k+n}| + \cdots = |R| \leq |a_{k+1}|$. Therefore, $|L - (a_1 + a_2 + a_3 + \cdots + a_k)| = |R| \leq |a_{n+1}|$, which is the desired conclusion.

For example, in the first of the three preceding examples, it was shown by the Ratio Test that the series

$$e^x = 1 + x + \frac{x^2}{2!} + \frac{x^3}{3!} + \cdots + \frac{x^n}{n!} + \cdots$$

converges for all real numbers x. Whenever $x < 0$, this series is alternating. For example, in the event $x = -1$ we have

$$e^{-1} = 1 - 1 + \frac{1}{2!} - \frac{1}{3!} + \frac{1}{4!} - \frac{1}{5!} + \cdots + \frac{(-1)^n}{n!} + \cdots$$

The first six terms of this series are $0, \frac{1}{2}, \frac{1}{3}, \frac{3}{8}, \frac{11}{30}$, and $\frac{53}{144} \approx .3680$. According to Theorem 7-9, the actual sum, $\frac{1}{e}$, differs from .3680 by no more than $\frac{1}{7!} = \frac{1}{5040} \approx .0002$. Thus $.3678 \leq \frac{1}{e} \leq .3682$, so that $\frac{1}{e} = .368$, correct to three decimal places.

EXERCISES

1. Investigate each series for convergence or divergence. Apply the Test for Alternating Series or the Ratio Test whenever possible. Where not possible, use any other means at your disposal. Justify your answers.

(a) $\frac{1}{5} - \frac{2}{5^2} + \frac{3}{5^3} - \frac{4}{5^4} + \cdots + \frac{(-1)^{n-1}n}{5^n} + \cdots$

(b) $\frac{1}{5} + \frac{2}{5^2} + \frac{3}{5^3} + \frac{4}{5^4} + \cdots + \frac{n}{5^n} + \cdots$

(c) $1 + \frac{1\cdot 3}{1\cdot 4} + \frac{1\cdot 3\cdot 5}{1\cdot 4\cdot 7} + \frac{1\cdot 3\cdot 5\cdot 7}{1\cdot 4\cdot 7\cdot 10} + \cdots + \frac{1\cdot 3\cdot 5\cdot 7\cdots(2n-1)}{1\cdot 4\cdot 7\cdot 10\cdots(3n-2)} + \cdots$

(d) $2 + \frac{1}{2} + \frac{8}{27} + \frac{16}{64} + \cdots + \frac{2^n}{n^3} + \cdots$

(e) $\frac{1}{1\cdot 4} - \frac{1}{2\cdot 5} + \frac{1}{3\cdot 6} - \frac{1}{4\cdot 7} + \cdots + \frac{(-1)^{n-1}}{n(n+3)} + \cdots$

(f) $\frac{1}{2} - \frac{2}{3} + \frac{3}{4} - \frac{4}{5} + \cdots + \frac{(-1)^{n-1}n}{n+1} + \cdots$

(g) $1 - 1 + 1 - 1 + 1 - 1 + \cdots + (-1)^{n-1} + \cdots$

(h) $1 + 0 - 1 + 0 + \cdots + \sin \frac{n}{2}\pi + \cdots$

(i) $1 - \frac{1}{\sqrt{2}} + \frac{1}{\sqrt{3}} - \frac{1}{\sqrt{4}} + \cdots + \frac{(-1)^{n-1}}{\sqrt{n}} + \cdots$

2. Given: $\log_e(1 + x) = x - \frac{x^2}{2} + \frac{x^3}{3} - \frac{x^4}{4} + \cdots + \frac{(-1)^{n-1}x^n}{n} + \cdots$.

(a) Let $\{a_n\} = \left\{\frac{(-1)^{n-1}x^n}{n}\right\}$ for $0 < x \leq 1$, and show by the Domination Principle (discussed in Chapter 5) that $\{|a_n|\} \to 0$.
(b) Show by means of the Test for Alternating Series that this series converges for $0 < x \leq 1$.
(c) Find $\log_e (1\frac{1}{3})$, correct to two decimal places.

3. Show that the series

$$\sin x = x - \frac{x^3}{3!} + \frac{x^5}{5!} - \frac{x^7}{7!} + \cdots + \frac{(-1)^{n-1}x^{2n-1}}{(2n-1)!} + \cdots$$

converges for
(a) $x = 2$,
(b) x any real number.

4. Given: $\log_e \left[\frac{1+x}{1-x}\right] = 2\left[x + \frac{x^3}{3} + \frac{x^5}{5} + \frac{x^7}{7} + \cdots + \frac{x^{2n-1}}{2n-1} + \cdots\right]$

Show that this series converges for $-1 < x < 1$.

Appendix

A Discussion of the Real Number System

The Decimal Fractions

Let us imagine for the present that the only numbers in our possession are the decimal fractions. A decimal fraction (or terminating decimal) is a number which can be placed in either of the two forms

$$z + \frac{a_1}{10} + \frac{a_2}{10^2} + \cdots + \frac{a_n}{10^n} \text{ and } -z - \left(\frac{a_1}{10} + \frac{a_2}{10^2} + \cdots + \frac{a_n}{10^n}\right),$$

where z is zero or a positive integer, the a_i's are the digits 0, 1, 2, 3, $\cdots$, 9, and n is some specified positive integer. Decimal fractions are commonly designated by one of the forms $z.a_1a_2a_3 \cdots a_n$ and $-z.a_1a_2a_3 \cdots a_n$. Some particular decimal fractions are .235, 7.002, and -2.516. Every integer is a decimal fraction; for example, 2.0, 25.0. Infinite decimals, such as the repeating decimals $.6666\cdots$ and $.242424\cdots$, are *not* decimal fractions.

Clearly a fraction whose numerator and denominator are both integers is a decimal fraction if and only if its denominator is a divisor of some power of 10. For example, $\frac{2}{5}$, $\frac{1}{4}$ and $\frac{3}{8}$ are decimal fractions, since 5 is a divisor of 10, 4 is a divisor of 10^2, and 8 is a divisor of 10^3. Their decimal representations are, respectively, .4, .25, and .375. However, $\frac{1}{3}$, $\frac{5}{6}$, and $\frac{2}{11}$ are *not* decimal fractions, since their denominators are not divisors of some power of 10. To show in another way that these are not decimal fractions, we can divide the numerator by the denominator and assure ourselves that a finite decimal representation is impossible. Dividing 1 by 3 gives unending repetition of digits, $.333 \cdots$. The same is true in division of 5 by 6 ($.833\cdots$) and 2 by 11 ($.1818\cdots$). Thus the set of decimal fractions excludes a host of rational numbers (numbers which can be placed in the form $\frac{a}{b}$, where a and b are integers and $b \neq 0$).

Since any two decimal fractions can be expressed as ratios whose denominators are powers of ten, in addition and subtraction the larger of these two denominators serves as a common denominator. In multiplication the product of the two denominators is again a power of ten. Thus we can add, subtract, and multiply any two members of the set of decimal fractions and again obtain a member of the set.

Within the set of decimal fractions, we are not always able to divide one member of this set by another member. For example, we cannot divide 1 by 3, for there is no decimal fraction which can be multiplied by 3 to obtain 1. (Let us recall that the basic meaning of division is as follows: To divide x by y means to find a number z (in the same set) such that $x = y \cdot z$, provided such a number z exists.) Faced with this dilemma, we do the best we can by looking for some decimal fraction which, when multiplied by 3, will give an approximation of 1. But *which* decimal fraction shall we use? Among the 1-place decimals, .3 is the best choice. Among the 2-place decimals, .33 is the best choice. Among the 3-place decimals, .333 is the best choice, and so on. Furthermore, .33 is a better choice than .3, and .333 is a better choice than .33, and, in general, for the sequence with terms .3, .33, .333, $\cdots$ (general term: $.3 + .03 + .003 + \cdots + 3(.1)^n$), any particular term is a better choice than each of the preceding terms.

Therefore, although we cannot find a number within the set of decimal fractions which is the quotient of 1 divided by 3, we can find a *sequence* of decimal fractions whose terms approximate the desired quotient to any degree of accuracy. Similarly, dividing 2 by 11 produces the sequence with terms .18, .1818, .181818, $\cdots$ (general term: $.18 + .0018 + .000018 + \cdots + 18(.01)^n$). Every term of this sequence is a decimal fraction, yet the quotient of 2 divided by 11 is not a decimal fraction, since there is no decimal fraction which can be multiplied by 11 to give 2.

In conclusion, although we can add, subtract, and multiply any two decimal fractions and always obtain a decimal fraction, we cannot divide two decimal fractions and always obtain a decimal fraction as quotient. In every case, however, we can find a sequence of decimal fractions whose terms approximate the desired quotient to any degree of accuracy. Accordingly, there are many convergent sequences whose terms are decimal fractions but whose limits are not decimal fractions.

The Rational Numbers

A rational number is a number which can be placed in the form $\frac{a}{b}$, where a and b are integers and b is not 0. As is true with decimal fractions, we can add, subtract, and multiply any two rational numbers, and the result will always be a rational number. Furthermore, if we divide any rational number by any rational number (except 0), we obtain a rational number. That is, we can freely divide with rational numbers, whereas we cannot with decimal fractions.

Moreover, every decimal fraction is a rational number, but not vice versa. For example, the two decimal fractions 1.5 and .273 can be written as $\frac{15}{10}$ and $\frac{273}{1000}$, respectively, yet $\frac{1}{3}$ and $\frac{2}{11}$, as we have discussed, are rational numbers which cannot be written in decimal fraction form. However, any rational number which is *not* a decimal fraction is the limit of a sequence of decimal fractions. For example, $\frac{1}{3}$ is the limit of the sequence with terms .3, .33, .333, $\cdots$, and $\frac{2}{11}$ is the limit of the sequence with terms .18, .1818, .181818, $\cdots$. In addition, every decimal fraction is the limit of a constant sequence of decimal fractions (e.g., 2.2 is the limit of $\{a_n\} = \{2.2\}$). In other words, *every rational number is the limit of a sequence of decimal fractions.* But does every convergent sequence whose terms are decimal fractions have a rational number as its limit? Actually, this question is hardly one of idle curiosity. What we are really asking here is whether or not the set of rational numbers makes up for the "inadequacies" of the set of decimal fractions.

Suppose we were to rent Madison Square Garden in New York City and by some extraordinary means we could place in this building all possible convergent sequences whose terms are decimal fractions. This collection of sequences would consist of sequences of the following types:

(a) Sequences whose limit is a decimal fraction, such as those with terms

$$.9, .99, .999, \cdots \to 1; \qquad .24, .249, .2499, .24999, \cdots \to .25.$$

This includes any constant sequence of decimal fractions, for example, $\{a_n\} = \{3.27\} \to 3.27$.

(b) Sequences whose limit is a rational number which is not a decimal fraction, such as those with terms

$$.3, .33, .333, \cdots \to \tfrac{1}{3}; \qquad .18, .1818, .181818, \cdots \to \tfrac{2}{11};$$
$$.142857, .142857142857, .142857142857142857, \cdots \to \tfrac{1}{7}.$$

(c) Sequences whose limit is not a rational number. For example, let b_n be the *largest* nth place decimal fraction whose square is *less* than 2. The first four terms of $\{b_n\}$ are 1.4, 1.41, 1.414, and 1.4142. Let c_n be the *smallest* nth-place decimal fraction whose square is *greater* than 2. The first four terms of $\{c_n\}$ are 1.5, 1.42, 1.415, and 1.4143. The limit of the two sequences $\{b_n\}$ and $\{c_n\}$ is $\sqrt{2}$. There is a very common proof of the fact that there is no rational number whose square is 2—that is, that $\sqrt{2}$ is not a rational number. We will not present this proof here, since it appears in a number of high school textbooks. We also could prove in a similar manner that $\sqrt{3}$, $\sqrt{5}$, $\sqrt{6}$, and, in general, $\sqrt{p}$, where p is any natural number not a perfect square, are irrational. (There are, of course, other irrational numbers.) Each of these numbers can be shown to be the limit of various sequences of decimal fractions.

The conclusion pertinent to our discussion of the real numbers is that the answer is "No" to the question "Does every convergent sequence whose terms are decimal fractions have a rational number as its limit?" We have, in effect, also answered "No" to the question "Does every convergent sequence whose terms are rational numbers have a rational number as its limit?"

The Real Numbers

Let us return to our collection of sequences in Madison Square Garden and enlarge our collection so that we now have all possible convergent sequences whose terms are *rational numbers.* Suppose further that we find the limit of each of these sequences. *The set of limits of all such sequences is called the set of real numbers.* We have defined the real numbers in terms of limits!

The real numbers fall into two categories:

(a) *The rational numbers.* Every rational number is the limit of a constant sequence of rational numbers, as well as of many other sequences. For example, $\frac{2}{3}$ is the limit of the constant sequence $\{a_n\} = \{\frac{2}{3}\}$, and it is also the limit of the two sequences with terms .6, .66, .666, $\cdots$ and .7, .67, .667, $\cdots$, respectively, and of other sequences.

(b) *The irrational numbers.* An irrational number is a real number which is not rational — i.e., an irrational number is a real number which cannot be expressed in the form $\frac{a}{b}$, where a and b are integers and b is not 0. Some particular irrational numbers are $\sqrt{2}$, $\sqrt{3}$, $\sqrt{5}$, $\sqrt[3]{7}$, and e.

Of course, there are sequences some (or all) of whose terms are irrational numbers. For example, the first, fourth, ninth, sixteenth, etc., terms of $\{c_n\} = \left\{\frac{1}{\sqrt{n}}\right\}$ are rational numbers, and the remaining terms are irrational.

The Completeness Property

Every convergent sequence we have encountered up to now has as its limit a real number. That this is true of all convergent sequences with terms in the real number system is so widely recognized that it is stated in terms of a property of the real numbers called the Completeness Property.

COMPLETENESS PROPERTY OF THE REAL NUMBER SYSTEM

Every convergent sequence whose terms are real numbers has as its limit a real number.

To help you comprehend the significance of the Completeness Property, we shall discuss the relation of this property to the least upper bound axiom and an accompanying theorem.

In Chapter 3, the least upper bound axiom stated that if S is any set of real numbers which is bounded above, then of all the upper bounds of S, one of these is less than any other upper bound of S and is called the least upper bound for S. On the basis of this axiom, we were able to prove the following theorem (Theorem 3-4): If $\{a\}$ is a sequence of real numbers which is bounded above and everywhere increasing, then $\{a_n\}$ converges, and the limit of $\{a_n\}$ is its least upper bound.

Without the least upper bound axiom, this theorem could not have been proved. Furthermore, the least upper bound axiom is true for the real numbers, but not for the rational numbers or the decimal fractions. In fact, *the least upper bound axiom and the Completeness Property go hand in hand; whenever one of them is true for a given system of numbers, the other is also true*. So Theorem 3-4 is true for the real numbers, but not for the rational numbers and the decimal fractions. The following example stresses this point.

Example Consider once again the two sequences $\{b_n\}$ and $\{c_n\}$, for which b_n is the largest nth-place decimal whose square is less than 2, and c_n is the smallest nth-place decimal whose square is greater than 2.

The first four terms of $\{b_n\}$ are 1.4, 1.41, 1.414, 1.4142. The first four terms of $\{c_n\}$ are 1.5, 1.42, 1.415, 1.4143.

The sequence $\{b_n\}$ is everywhere increasing and is bounded above. (Every term of $\{c_n\}$ is an upper bound.) If we were working within the set of decimal fractions, or even within the set of rational numbers, this sequence would not have a least upper bound. (Try to find one.) Within the set of real numbers, however, the least upper bound for $\{b_n\}$ does exist, and it is the limit of the sequence, namely, $\sqrt{2}$.

Final Remarks

In this discussion we have given a precise meaning to the concept of a real number in terms of limits. Indeed, one of the many advantages of studying sequences and limits is the accompanying insight into the structure of our number systems.

In terms of limits, the irrational numbers make sense. They are simply some rather special limits of rational sequences. We cannot name these numbers using rational number form. Instead, we use various other names, such as $\sqrt{2}$, $(\sqrt{2/3})$, $\sqrt[3]{7}$, $-\sqrt{5}$, $\log_e 5$, $\sin 2$, $\tan(\pi/13)$. In calculations, of course, we use some particular term of a sequence whose terms are decimal fractions and whose limit is the given irrational number; which term we choose depends upon the degree of accuracy expected for the situation.

A Summary of Methods for Proving Convergence of Sequences

Listed for each method are corresponding pages of the text containing the statement of the principle and some examples.

1. Definition of Convergence (pp. 34–38)

A sequence $\{a_n\}$ is said to converge to a number A (or to approach A as a limit) if for every neighborhood of A there can be found a natural number M such that all terms of $\{a_n\}$ with $n \geq M$ are in the neighborhood.

2. Alternate Meaning of Convergence (pp. 39–40)

If $\{a_n\} \to A$ and $\langle A - E, A + E\rangle$ is *any* neighborhood of A, then at most a finite number of terms of $\{a_n\}$ lie outside $\langle A - E, A + E\rangle$. In other words, if there can be found *just one* neighborhood of A outside of which there are an infinite number of terms of the sequence, then the sequence would not converge to A.

3. Convergent Increasing Sequences (pp. 57–60, 62–68)

Theorem 3–4. If $\{a_n\}$ is a sequence, with all terms real numbers, such that $\{a_n\}$ is bounded above and everywhere increasing, then

(1) $\{a_n\}$ converges, and

(2) the limit of $\{a_n\}$ is its least upper bound.

Theorem 3–8. If $\{a_n\}$ is a sequence which is bounded above and which is increasing beginning with some term other than the first term, then $\{a_n\}$ converges. (Its limit might or might not be its least upper bound.)

4. Convergent Decreasing Sequences (pp. 60–68)

Theorem 3–7. If $\{a_n\}$ is a sequence, with all terms real numbers, such that $\{a_n\}$ is bounded below and everywhere decreasing, then $\{a_n\}$ converges, and the limit of $\{a_n\}$ is its greatest lower bound.

Theorem 3–9. If $\{a_n\}$ is a sequence which is bounded below and which is decreasing beginning with some term other than the first term, then $\{a_n\}$ converges. (Its limit might or might not be its greatest lower bound.)

5. Geometric Sequences

(pp. 68–69)

A geometric sequence $\{a_1 r^{n-1}\}$

(1) converges to 0 if $0 < |r| < 1$;

(2) converges to a_1 if $r = 1$; and

(3) diverges otherwise.

6. Sums, Differences, Products, and Quotients of Sequences

(pp. 106–110, 115–117)

Theorems 5–1, 5–2, 5–3

If $\{a_n\} \to A$ and $\{b_n\} \to B$, then $\{a_n + b_n\} \to A + B$, $\{a_n - b_n\} \to A - B$, and $\{a_n b_n\} \to AB$.

Theorem 5–4. If $\{a_n\} \to A$ and $\{b_n\} \to B$ $(B \neq 0)$ and if no term of $\{b_n\}$ is 0, then the quotient sequence $\left\{\frac{a_n}{b_n}\right\}$ converges to $\frac{A}{B}$.

7. Domination Principle

(pp. 118–119)

Theorem 5–5. If two sequences $\{a_n\}$ and $\{b_n\}$ both converge to the same number L and if $\{c_n\}$ is a sequence such that $a_n \leq c_n \leq b_n$ for all n greater than or equal to some natural number M, then $\{c_n\}$ also converges to L.

8. Geometric Sequences of Partial Sums

(pp. 121, 123–126)

A geometric sequence of partial sums

$$\{S_n\} = \{a_1 + a_1 r + a_1 r^2 + \cdots + a_1 r^{n-1}\} \qquad (r \neq 1)$$

equals $\left\{a_1 \left(\frac{1 - r^n}{1 - r}\right)\right\}$ and converges to $\frac{a_1}{1 - r}$ if and only if $0 < |r| < 1$.

Answers to Selected Problems

Section 1-2, page 7

1. (c). **3.** (a). **5.** (f). **7.** (i).

Section 1-3, pages 9–10

2. a_n: $1, -1, 1, -1, 1$;
c_n: $\frac{1}{3}, \frac{1}{4}, \frac{1}{5}, \frac{1}{6}, \frac{1}{7}$;
e_n: $\frac{1}{3}, \frac{1}{6}, \frac{1}{11}, \frac{1}{18}, \frac{1}{27}$;
h_n: $-1, \frac{1}{2}, -\frac{1}{3}, \frac{1}{4}, -\frac{1}{5}$;
k_n: $2, \frac{1}{2}, 8, \frac{1}{8}, 32$;
q_n: $0, 2, 0, 2, 0$;
s_n: $1, \frac{1}{2}, \frac{1}{3}, \frac{1}{10}, \frac{1}{29}$.

3. (a) $\{j_n\}$. (c) $\{q_n\}$. (e) $\{r_n\}$. (g) $\{p_n\}$.

Section 1-4, pages 12–13

1. (b), (f), (m). **3.** (1). **5.** (d), (k).

7. Possible solutions:

$$f_n = 2^n + (n-1)(n-2)(n-3)(n-4);$$
$$f_n = 2^n - (n-1)(n-2)(n-3)(n-4).$$

9. Possible solutions:

$$s_n = 2^n - 1;$$
$$s_n = 2^n - 1 + (n-1)(n-2)(n-3)(n-4).$$

10. Possible solutions:

$$a_n = \begin{cases} (n+1)^2 & \text{for } 1 \leq n \leq 4 \\ n & \text{otherwise,} \end{cases}$$
$$b_n = \begin{cases} (n+1)^2 & \text{for } 1 \leq n \leq 4 \\ n-1 & \text{otherwise;} \end{cases} \quad \text{or}$$
$$b_n = (n+1)^2.$$

Section 1-5, page 16

1. 1, 2, 3, 6, 12. **3.** 1, 2, 3, 6, 12. **5.** 1, 2, 3, 7, 46. **6.** 1, 3.

8. Let $b_1 = 1$, and for $n \geq 2$, let $b_n = nb_{n-1}$.

9. $a_n = 3 + 12(n - 1)$ or $12n - 9$.

11. $c_n = (-1)^{n+1}5$. **13.** $f_n = -5 - 2(n - 1)$ or $-2n - 3$.

14. $a_1 = 8$, and for $n \geq 2$, $a_n = a_{n-1} + 3$.

17. $a_3 = p + 2(q - p)$ or $2q - p$; $a_n = p + (n - 1)(q - p)$.

Section 2-3, page 24

1. (a) $n > 5$. (c) $n < 35$. (e) $n > 8$. (g) $n > 197$.

2. (a) All a_n with $n > 245$. (c) all c_n with $n > 3$.

Section 2-4, page 27

1. (a) $n > \frac{1}{E}$. (c) $n > \sqrt{\frac{13}{E}}$.

2. (a) All a_n with $n > \frac{3}{4E} - \frac{3}{2}$; all a_n; all a_n with $n \geq 2$.

(c) All d_n with $n > \sqrt{\frac{5}{E}}$; all d_n with $n \geq 4$; all d_n with $n \geq 4$; all d_n with $n \geq 5$.

3. (a) All terms. (c) first 8 terms. (e) all c_n with $n > \frac{3}{4E} + \frac{1}{2}$.

4. (a) All terms. (c) all b_n with $n > 13$.

(e) All b_n with $n > \frac{13}{3E}$. (g) all b_n with $n > 26$.

5. (a) All a_n with $n > 199$.

Section 2-5, pages 33–34

1. Limit is $\frac{1}{2}$. For $n > 5$, a_n is in $\langle .4, .6\rangle$; for $n > 50$, a_n is in $\langle .49, .51\rangle$; for $n > \frac{1}{2E}$, a_n is in $\langle \frac{1}{2} - E, \frac{1}{2} + E\rangle$.

3. (a) All c_n with $n > 3$.

Section 2-6, pages 38–39

5. $\{a_n\} \to -1\frac{1}{2}$; $\{b_n\} \to \frac{2}{5}$; $\{c_n\} \to 1$;
$\{d_n\} \to 0$; $\{f_n\} \to 2$; $\{g_n\} \to 0$; $\{h_n\} \to 1$.

11. (a) All q_n with $n \geq 3$. **12.** (b) All a_n with $8 \leq n \leq 16$. (c) .9.

Section 2-7, page 44

1. $a_n < .01 \leftrightarrow 500 < (n+3)^2$, which is true whenever $500 < n^2$; so choose $M \geq 23$ (actually any $M \geq 20$ will do).

3. $c_n < .04 \leftrightarrow 50 < n^2 + 2^n$, which is true whenever $50 < n^2$; so choose $M \geq 8$. Or, since $50 < n^2 + 2^n$ is true also whenever $50 < 2^n$, choose $M \geq 6$ (actually any $M \geq 5$ will do).

6. $g_n < .03 \leftrightarrow 100\sqrt{n} < 3(n+1)$, which is true whenever $100\sqrt{n} < 3n \leftrightarrow \sqrt{n} > 33\frac{1}{3}$; so choose $M \geq 34^2 = 1156$ (actually any $M \geq 1111$ will do).

Section 2-8, page 48

1. $\{a_n\}$ diverges by Theorem 2-1.

3. $\{d_n\}$ diverges by comparison with $\{n\}$, since $n \leq n^2 - n \leftrightarrow n \geq 2$.

5. $\{g_n\} \rightarrow -3$.

7. $\{j_n\}$ diverges by comparison with $\{n\}$, since $n^{n-5} \geq n \leftrightarrow n^n \geq n^6 \leftrightarrow n \geq 6$.

9. $\{p_n\}$ diverges by Theorem 2-1. **11.** $\{r_n\} \rightarrow 2$.

Section 3-1, page 52

1. Not everywhere increasing. **3.** Everywhere increasing.

5. Not everywhere increasing. **7.** Everywhere increasing.

Section 3-2, page 57

1. $\{a_n\}$ increases without bound by comparison with $\{e_n\} = \{\sqrt{n}\}$.
$\{c_n\}$ increases without bound by comparison with $\{g_n\} = \{n^3\}$.

2. (a) 3, 6, and 8 increase without bound; none are convergent.

Section 3-4, page 61

1. $\{a_n\} \rightarrow 2$, Theorem 3-4; $\{c_n\}$ is neither increasing nor decreasing;
$\{g_n\} \rightarrow 0$, Theorem 3-7; $\{k_n\} \rightarrow 1$, Theorem 3-7.

Section 3-5, pages 67–68

(Graphs are not included here but should be included in your work.)

1. $\{a_n\} = \{n^2 - 10n\}$ is decreasing for $1 \leq n \leq 4$ and increasing for $n \geq 5$. The sequence is unbounded above and hence divergent. A lower bound is $a_5 = -25$ (any number ≤ -25 will do).

3. $\{c_n\} \to 1$, is decreasing for $1 \leq n \leq 14$ and increasing for $n \geq 14$. A lower bound is 0; an upper bound is 4 (any number ≥ 4 will do).

5. $\{f_n\} \to 0$, is increasing for $1 \leq n \leq 2$ and decreasing for $n \geq 3$. A lower bound is -5 (any number ≤ -5 will do), and an upper bound is $\frac{13}{9}$ (any number $\geq \frac{13}{9}$ will do).

7. $\{h_n\} \to 1$, is increasing for $1 \leq n \leq 10$ and decreasing for $n \geq 11$. A lower bound is $\frac{1}{4}$ (any number $\leq \frac{1}{4}$ will do), and an upper bound is $h_{11} = 1\frac{9}{68}$.

11. 3, 5, 6, 7.

13. $\{t_n\} = \left\{\dfrac{1}{\sqrt{n+1} - \sqrt{n}} \cdot \dfrac{\sqrt{n+1} + \sqrt{n}}{\sqrt{n+1} + \sqrt{n}}\right\} = \{\sqrt{n+1} + \sqrt{n}\}$ increases without bound by comparison with $\{\sqrt{n}\}$.

15. Any constant sequence is both everywhere increasing and everywhere decreasing.

Section 3-6, page 69

1. (a) $r = 1$ or 3. (b) $r = \frac{1}{2}$ or 1. (c) $r = -\frac{1}{2}, \frac{1}{2}$, or 1. **3.** (a) h. (b) f.

Section 4-1, pages 78–79

1. (a) $f(x) = 2x - 3$. (c) $f(x) = -x + 11$. (e) $f(x) = -2x + 3$.

2. (a) $f(x) = 4x - 2$. (c) $f(x) = 5x + 7$.

3. 49. $-\frac{7}{170}$; $-\frac{2}{65}$. 11; $9 + \dfrac{1}{n}$. -3; $1 + \dfrac{1}{n} \cdot \frac{1}{10}$; $\dfrac{1}{16 + \dfrac{2}{n}}$.

Section 4-2, pages 84–85

1. (a) 90, 70, 10, 10, 70, 90 (feet per second).

2. (a) 4.16, 5.08, 4.80, 6.82, 8.27 (dollars per year).

Section 4-3, pages 92–94

1. (a) 20, 32, 36, 32, 20, 0 (feet). (b) 20, 12, 4, -4, -12, -20 (ft./sec.).

(c) $\{a_n\} = \left\{8 - \dfrac{4}{n}\right\} \to 8$.

3. (a) 1, 3, 5, 7 (units). (b) $\left\{6 + \dfrac{1}{n}\right\} \to 6$.

4. (a) $\frac{3}{4}$ grams/hr. (b) $\frac{1}{6}$ grams/hr.; $\frac{9}{52}$ grams/hr.

(c) $$\left\{-\frac{f\left(4+\frac{1}{n}\right)-f(4)}{\frac{1}{n}}\right\} = \left\{-\frac{-3}{4\left(4+\frac{1}{n}\right)}\right\} = \left\{-\frac{-3}{16+\frac{4}{n}}\right\} \to \frac{3}{16} \text{ grams/hr.}$$

5. (a) $f(5) = 200$ in.

Section 4-4, page 98

1. (a) $y = 3x - 2$. (c) $y = 0$. **2.** (a) $y = 2x - 2$.

3. (a) $y = 2x - 3$. (c) $y = -2x + 1$.

Section 4-5, page 104

1. (a) 5. (b) $2c + 1$. (c) $(-\frac{1}{2}, -\frac{1}{4})$. (d) $x < -\frac{1}{2}$.

3. (a) -1. (c) $y = -\frac{1}{4}x + 1$; $y = -9x + 6$.

4. (a) $2c$. **5.** (a) $\frac{\pi}{2}, \frac{3\pi}{2}$.

Section 5-1, pages 110–111

1. $\{a_n\} \to \frac{6}{13}$; $\{c_n\} \to -\frac{1}{2}$; $\{f_n\} \to \frac{5}{2}$; $\{h_n\} \to 0$; $\{p_n\} \to 8$; $\{r_n\} \to 10$; $\{t_n\} \to 54$; $\{w_n\} \to 81$. **2.** (a) $\frac{1}{60}$. (b) $\frac{1}{24}$.

Section 5-3, pages 117–118

1. $\{a_n\} \to 3$; $\{c_n\}$ diverges; $\{e_n\}$ diverges; $\{g_n\} \to 2$; $\{j_n\} \to 1$; $\{p_n\}$ diverges; $\{r_n\} \to \frac{1}{2}$; $\{t_n\} \to \frac{1}{2}$; $\{v_n\} \to 24$.

2. (a) 4.1, 4.17, 4.181, 4.1814; limit is $\sqrt{3} + \sqrt{6}$. (c) 1.4118, 1.4104, 1.4140, 1.4142 (approximately); limit is $\sqrt{2}$.

Section 5-4, page 119

1. $\{a_n\} \to 0$ by the Domination Principle; $\{c_n\} \to 0$ by the Domination Principle, since $-\frac{1}{n} \le \frac{\sin n}{n} \le \frac{1}{n} \leftrightarrow -1 \le \sin n \le 1$, which is true for all natural numbers; $\{f_n\} \to 0$ by the Domination Principle; $\{h_n\} \to 1$; $\{p_n\} \to 3$.

Section 6-1, pages 124–126

1. $\{S_n\} \to 49\frac{49}{99}$; $\{T_n\} \to \frac{7}{9}$; $\{V_n\}$ diverges; $\{W_n\} \to \frac{2}{3}$; $\{P_n\} \to 3$.

2. (a) $\{96 + 96(\frac{3}{4}) + 96(\frac{3}{4})^2 + \cdots + 96(\frac{3}{4})^{n-1}\} \to 384$ ft. (b) iv.

6. Limit does not exist. **7.** 24 in.

Section 6-2, page 132

1. $$U_n = \frac{1}{n}\left(\frac{1}{n}\right)^3 + \frac{1}{n}\left(\frac{2}{n}\right)^3 + \frac{1}{n}\left(\frac{3}{n}\right)^3 + \cdots + \frac{1}{n}\left(\frac{n}{n}\right)^3$$
$$= \frac{1}{n^4}[1^3 + 2^3 + 3^3 + \cdots + n^3] = \frac{1}{n^4}\left[\frac{n(n+1)}{2}\right]^2 = \frac{1}{4}\left(1 + \frac{1}{n}\right)^2;$$
limit is $\frac{1}{4}$ square unit.

3. $\frac{1}{2}$ square unit. **5.** (a) $2\frac{2}{3}$ square units. (b) $2\frac{1}{3}$ square units.

Section 6-3, page 136

1. (a) $$V_n = \frac{2}{n}\pi\left[\frac{3}{2}\left(\frac{2}{n}\right)\right]^2 + \frac{2}{n}\pi\left[\frac{3}{2}\left(\frac{4}{n}\right)\right]^2 + \frac{2}{n}\pi\left[\frac{3}{2}\left(\frac{6}{n}\right)\right]^2 + \cdots + \frac{2}{n}\pi\left[\frac{3}{2}\left(\frac{2n}{n}\right)\right]^2.$$

(b) $V_n = \frac{18}{n^3}\pi[1^2 + 2^2 + 3^2 + \cdots + n^2] = 3\pi\left(1 + \frac{1}{n}\right)\left(2 + \frac{1}{n}\right)$; limit is 6π.

3. (a) $33\frac{1}{3}\pi$ cubic inches.

Section 6-4, page 141

1. (a) $\frac{1}{2}\pi$ cubic units. (b) 39π cubic units. (c) 6π cubic units.

Section 7-2, pages 156–157

1. (a) $\frac{2}{3}$. (c) No limit, since this is a geometric series with $r > 1$.

2. (a) $.7 + .07 + .007 + \cdots + .7(.1)^{n-1} + \cdots = \frac{7}{9}$. (c) The geometric series with $a = .3$ and $r = -.1$ converges to $\frac{.3}{1 - (-.1)} = \frac{.3}{1.1} = \frac{3}{11}$.

6. $$\log_e 9 = \log_e\left[\frac{1 + .8}{1 - .8}\right] = 2\left[.8 + \frac{(.8)^3}{3} + \frac{(.8)^5}{5} + \frac{(.8)^7}{7} + \cdots\right].$$

$S_1 = 1.6$, $S_2 = 1.9413$ (approximately).

Section 7-4, pages 163–164

2. Series (b).

4. (a) Diverges, by comparison with $1 + \frac{1}{2} + \frac{1}{3} + \cdots + \frac{1}{n} + \cdots$.

(c) Diverges, since $\left\{\frac{n}{3n+1}\right\}$ does not converge to 0.

(e) Converges, by comparison with $1 + \frac{2}{3} + \frac{4}{9} + \cdots + (\frac{2}{3})^{n-1} + \cdots$.

(g) Diverges, by Theorem 7-6, with $c = \frac{1}{12}$.

(i) Diverges, by comparison with $1 + \frac{1}{2} + \frac{1}{3} + \cdots + \frac{1}{n} + \cdots$.

(k) This series equals the sequence $\{a_n\} = \left\{\frac{n}{n+1}\right\}$, and hence converges to 1.

5. Diverges, by comparison with $1 + \frac{1}{2} + \frac{1}{3} + \cdots + \frac{1}{n} + \cdots$.

Section 7-5, pages 170–171

(TAS = Test for Alternating Series; RT = Ratio Test)

1. (a) Converges, by either TAS or RT.

(c) Converges by RT: $\left\{\left|\frac{a_{n+1}}{a_n}\right|\right\} = \left\{\frac{2n+1}{3n+1}\right\} \to \frac{2}{3}$.

(e) Converges by TAS. (g) Diverges, since $\{|a_n|\} = \{1\}$ does not converge to 0.
(i) Converges by TAS.

• *Glossary*

Alternating Series. $a_1 + a_2 + a_3 + \ldots + a_n + \ldots$ is called an *alternating series* if the terms of $\{a_n\}$ are alternately positive and negative; it is convergent if $\{|a_n|\}$ converges to 0 and is everywhere decreasing.

Arithmetic Sequence. A sequence in which the difference between any term (after the first term) and its predecessor is always the same. Can be described by the general term $a_1 + d(n - 1)$, where a_1 is the first term and d is the common difference; the numbers a_1 and d may be positive or negative.

Average Rate of Change of a Function f in Interval $[a,b]$. The number $\frac{f(b) - f(a)}{b - a}$. In particular, this number is called the *average velocity* of f in $[a,b]$ if $f(t)$ describes the position at time t of an object moving in a straight line.

Boundedness. A sequence is said to be *bounded above* if there is a real number B (called an *upper bound* for the sequence) such that no term of the sequence is greater than B; if no such number B exists, the sequence is said to be *unbounded above.* [The terms *bounded below*, *lower bound* and *unbounded below* are defined similarly, by replacing "is greater than B" by "is less than B."]

Comparison Test (for infinite series). If $\{a_n\}$ and $\{b_n\}$ are two sequences such that for all natural numbers n, $0 < a_n \leq b_n$, then (i) if $a_1 + a_2 + a_3 + \ldots + a_n + \ldots$ diverges, $b_1 + b_2 + b_3 + \ldots + b_n + \ldots$ also diverges; (ii) if $b_1 + b_2 + b_3 + \ldots + b_n + \ldots$ converges, $a_1 + a_2 + a_3 + \ldots + a_n + \ldots$ also converges.

Comparison Principle (for sequences). A method for showing that a sequence increases without bound (and hence diverges). Let $\{a_n\}$ be a sequence which increases without bound and let $\{b_n\}$ be a sequence which is everywhere increasing. If there is a particular natural number M such that, for all natural numbers $n \geq M$, $b_n \geq a_n$, then $\{b_n\}$ also increases without bound. [A similar principle holds for decreasing sequences.]

Constant Sequence. A sequence having the same number for each term.

Convergence. If the number A is the limit of a sequence, then the sequence is said to "converge to A". If a sequence converges (i.e., has a limit), then it is said to be a *convergent sequence.*

Decreasing Sequences. If no term is less than the succeeding term, a sequence is said to be *everywhere decreasing*. If a sequence is everywhere decreasing and is unbounded below, then the sequence is said to *decrease without bound*.

Divergent Sequence. A sequence which does not converge — i.e., which does not have a limit.

Domination Principle. If two sequences $\{a_n\}$ and $\{b_n\}$ both converge to the same number L and if $\{c_n\}$ is a sequence such that $a_n \leq c_n \leq b_n$ for all n greater than or equal to some natural number M, then $\{c_n\}$ also converges to L.

e. The limit of $\left\{\left(1 + \frac{1}{n}\right)^n\right\}$ and the base for the *natural logarithms*.

Fibonacci Sequence. The sequence $\{a_n\}$ such that $a_1 = 1$, $a_2 = 1$, and for $n \geq 3$, $a_n = a_{n-2} + a_{n-1}$ (i.e., any term after the second term is the sum of the two preceding terms).

Function. A set of ordered pairs of real numbers in which no two pairs have the same first elements and different second elements.

Geometric Series. The infinite series $a + ar + ar^2 + ar^3 + ... + ar^{n-1} + ...$, which has the sum $\frac{a}{1 - r}$ if and only if $0 < |r| < 1$, and diverges if and only if $|r| \geq 1$.

Geometric Sequence. A sequence in which the ratio of any term (after the first term) to its predecessor is always the same. Can be described by the general term a_1r^{n-1} where a_1 is the first term and r is the common ratio; the numbers a_1 and r may be positive or negative. Converges if and only if $-1 < r < 0$ or $0 < r \leq 1$.

Geometric Sequence of Partial Sums. $\{a + ar + ar^2 + ... + ar^{n-1}\}$ which equals $\left\{a\left(\frac{1 - r^n}{1 - r}\right)\right\}$

Greatest Lower Bound Axiom. Let S be any set of real numbers (not necessarily the terms of a sequence) which is bounded below. Of all the lower bounds of set S, one of these is greater than any other lower bound of S and is called the *greatest lower bound* for S.

Increasing Sequences. If no term is greater than the succeeding term, a sequence is said to be *everywhere increasing*. If a sequence is everywhere increasing and is unbounded above, then the sequence is said to *increase without bound*.

Infinite Series. The expression $a_1 + a_2 + a_3 + ... + a_n + ...$, for any sequence $\{a_n\}$. If the corresponding sequence of partial sums, $\{S_n\} = \{a_1 + a_2 + a_3 + ... + a_n\}$ converges to L, then $a_1 + a_2 + a_3 + ... + a_n ...$ is simply another name for L and is called the "sum" of the infinite series.

Instantaneous Rate of Change (of a function at c). Limit of $\left\{\frac{f(c + a_n) - f(c)}{a_n}\right\}$ provided this limit exists for all sequences $\{a_n\} \to 0$.

Interval: Closed. The set of points in the line segment whose end points are a and b, with a and b included.

Interval: Open. The set of points in the line segment whose end points are a and b, with a and b excluded.

Least Upper Bound Axiom. Let S be any set of real numbers (not necessarily the terms of a sequence) which is bounded above. Of all the upper bounds of set S, one of these is less than any other upper bound of S and is called the *least upper bound* for S.

Limit of a Sequence. By definition, a sequence $\{a_n\}$ is said to *approach the number A as a limit* (or *to converge to A*) if for every neighborhood of A there can be found a natural number M such that all terms of $\{a_n\}$ with $n \geq M$ are in the neighborhood. The number A is called *the limit of the sequence*. The notation $\{a_n\} \rightarrow A$ is read, "The sequence $\{a_n\}$ approaches A as a limit," or "The sequence $\{a_n\}$ converges to A."

Neighborhood. For any point r on a real number line, an open interval having r as midpoint is called a *neighborhood of r*. The positive number d which represents the distance from r to either of the end points of the neighborhood is called the *radius* of the neighborhood.

Ratio Test. A means for determining convergence or divergence for a series $a_1 + a_2 + a_3 + ... + a_n + ...$ by investigating convergence of the sequence $\left\{\left|\frac{a_{n+1}}{a_n}\right|\right\}$.

Recursive Sequence. The first R terms are stated, for some natural number R, and each succeeding term is defined as a function of one or more of the preceding terms.

Sequence. A sequence $\{b_n\}$ is a function which associates with every natural number n one and only one real number b_n. The numbers b_n are called the *terms* of the sequence.

Sequence of Partial Sums. For every sequence $\{a_n\}$, there is a *sequence of partial sums corresponding to* $\{a_n\}$ with general term $S_n = a_1 + a_2 + a_3 + ... + a_n$.

Subsequence. A sequence $\{b_n\}$ is called a *subsequence* of $\{a_n\}$ provided: (i) every term of $\{b_n\}$ is also a term of $\{a_n\}$, and (ii) the same order is preserved — i.e., if b_p and b_q are terms of $\{b_n\}$ with $p < q$, and if $b_p = a_r$ and $b_q = a_s$, then we must have $r < s$.

Tangent to a Curve. The tangent to the graph of a function f at point $(c, f(c))$ is the line which passes through this point and whose slope is the instantaneous rate of change of f at c, if the rate of change exists.

• *Index*